ADVANCED MATRIX THEORY FOR SCIENTISTS AND ENGINEERS

Qu'on ne dise pas que je n'ai rien dit de nouveau: la disposition des matières est nouvelle; quand on joue à la paume, c'est une même balle dont joue l'un et l'autre, mais l'un la place mieux

Pascal

ADVANCED MATRIX THEORY FOR SCIENTISTS AND ENGINEERS

Assem S. Deif

Associate Professor of Mathematics,
Faculty of Engineering, Cairo University

ABACUS PRESS, TUNBRIDGE WELLS & LONDON

HALSTED PRESS DIVISION
JOHN WILEY & SONS, NEW YORK & TORONTO

First published in 1982 by

ABACUS PRESS

Abacus House,
Speldhurst Road,
Tunbridge Wells,
Kent,
TN4 0HU,
England

British Library Cataloguing in Publication Data

Deif, Assem
Advanced matrix theory for scientists
and engineers.
1. Matrices
I. Title
512.9′434 QA188

ISBN 0-85626-327-3 Abacus Press
ISBN 0-470-27316-1 Halsted Press

Typeset and printed in Great Britain by
Pintail Studios Ltd
Ringwood, Hampshire

TO MY PARENTS

Bella & Shawki

PREFACE

This course in matrix theory is the result of a series of lectures given by me for many years to advanced undergraduate and first year postgraduate engineering students at Cairo University. It is natural that the book cannot cover all aspects of the theory. However I have tried to furnish the reader with the main outline of it. I do not pretend that the topics discussed are new, rather that the exposition of the subject-matter is new. The book is more like a mixture of something old and something new, something borrowed and something blue!

Among the pioneering works on matrix theory are the books by Frazer/Duncan/Collar, Gantmacher, Bellman and Pease. This book is not meant to compete with them; but the available books on the subject either explore the subject in a deep theoretical manner and are aimed at a reader with a strong mathematical background or are introductory texts which do not delve far into the subject. It was my observation that there is a need for a third kind of book which presents advanced topics in a manner accessible to scientists and engineers and with an applied problem bias. In this book, my aim has been to give a simple account of the subject, focusing on related applications, without drilling deep into the subject and at the same time without lacking rigour.

Matrix theory deserves to be called, as Bellman quotes, the arithmetic of higher mathematics. Almost every advanced engineering application involves matrices. The reason is that to deal with complex systems of many items necessitates the existence of a mathematical tool capable of relating together many items. Matrix methods are quite able to do that. Relevant applications of matrix theory exist in electrical networks, the theory of structures, mechanical systems, economic studies, etc.

In writing this book, I have tried to make it intelligible to many readers, who need not be equipped with much mathematical background to be able to digest it. As the pages unfold, the reader will soon become aware of the many important applications of matrix theory. Indeed, matrix theory lies at the core of applied mathematical techniques; and for an engineer a course in the subject is not wasted. Also, to make it consistent with first courses in mathematics I have tried to treat such courses in a generalized matrix form, even to show elegance when establishing explicit formulae. And in order to help the reader assimilate the subject-matter, many examples have been included in the text. Also there are many problems which are borrowed from other sources or are original creations, to let the reader test his capabilities in problem solving: all these are assembled at the end of sections. Some of them are not simple; should the reader be unable to solve them, he should not be disappointed, since most of the problems contain complementary theorems which could not be included in the text. Still the skill needed to solve them is by no means outside the scope of most readers of the book.

The book consists of six chapters. The first five are common to almost all books on matrices, while the sixth is sometimes missing from such works. I have tried in this chapter to brief the reader on perturbation techniques applied to important matrix functions. I believe that such a subject has many echoes in different areas of applied mathematics.

And since the book could find use by many interested readers it seemed wise to deduce all the results on general matrices. This does not mean that matrices of special interest have not received their share of the theory; some of their properties are given as problems for the reader.

Finally, I should like to thank Dr. F. Smith, for reading the whole manuscript and for valuable suggestions. My sincere thanks also go to Mr. M. Ziada for typing a difficult manuscript.

ASSEM DEIF

November 1981

CONTENTS

CHAPTER ONE

BASIC DEFINITIONS

1.1. Introduction

A matrix is a rectangular array of scalars. The latter may be real or complex. The rectangular array takes the general form

$$A_{m,n} = \begin{bmatrix} a_{11} & a_{12} & \cdots & a_{1n} \\ a_{21} & a_{22} & \cdots & a_{2n} \\ \vdots & & & \\ a_{m1} & & \cdots & a_{mn} \end{bmatrix}$$

The matrix A has m rows and n columns. Any element in A has the representation a_{ij} where the subscript i refers to the row and j to the column where the element lies. A matrix x of one row only or of one column only i.e.

$$x = [x_1 x_2 \ldots x_n] \quad \text{or} \quad x = \begin{bmatrix} x_1 \\ x_2 \\ \vdots \\ x_n \end{bmatrix}$$

is called a vector. Hence each row in A is a vector as well as each column. The number of elements in the vector is equal to the dimension of the space where the vector lies. If the elements of x are real, the vector x is defined in the real Euclidean space $\mathbb{R}^n$. Instead, if one or more elements in x are complex, the vector will be defined in the complex Euclidean space $\mathbb{C}^n$. Therefore the matrix A can be seen as an array of n vectors in an m-dimensional space or m vectors in an n-dimensional space. If the elements of A are real, its vectors are defined in the real Euclidean space, and if one element of A or more are complex, the vectors of A will consequently be defined in the complex Euclidean space. If m equals n, the matrix A is called a square matrix; and it follows that the number of vectors in A is equal to the dimension of the space. Therefore the matrix A has an order equal to n.

Example. Let

$$A = \begin{bmatrix} 1 & 2 & -3 \\ 0 & 1 & 4 \end{bmatrix}, \quad x = \begin{bmatrix} 1+i \\ -1 \\ 3 \end{bmatrix}, \quad i^2 = -1$$

Then $x \in \mathbb{C}^3$ and A consists of three column vectors in $\mathbb{R}^2$ or two row vectors in $\mathbb{R}^3$.

Two matrices A and B are said to be equal if their corresponding elements are equal, i.e. if

$$a_{ij} = b_{ij}, \quad \text{for all } i = 1, \ldots, m \text{ and } j = 1, \ldots, n$$

Obviously A and B can be equal only when they have the same size, i.e. the same number of rows and columns.

1.2. Common types of matrix

Although a matrix A takes the general form illustrated above, in most physical applications we frequently meet some special forms of matrix like the following:

1. *Diagonal matrix D.* This is a square matrix whose off-diagonal elements are zero, i.e.

$$a_{ij} = 0, \quad \text{for all } i \neq j.$$

This does not mean that $a_{ii} \neq 0$

Example

$$D_5 = \begin{bmatrix} -3 & & & & \\ & 1 & & & \\ & & 0 & & \\ & & & 5 & \\ & & & & 2+i \end{bmatrix}, \quad i^2 = -1$$

2. *Unit matrix I.* This is a diagonal matrix where $a_{ii} = 1$, for all i.

Example

$$I_3 = \begin{bmatrix} 1 & & \\ & 1 & \\ & & 1 \end{bmatrix}$$

3. *Zero matrix O.* Here each element is zero, i.e. $a_{ij} = 0$, for all i, j.

Example

$$O_{2,3} = \begin{bmatrix} 0 & 0 & 0 \\ 0 & 0 & 0 \end{bmatrix}$$

4. *Echelon form of matrix.* This has the following properties

(a) The first non-zero element in each row is one.

(b) In each row after the first, the number of zeros preceding the first non-zero element exceeds the number of zeros preceding the first non-zero element in the previous row.
(c) When the first non-zero element in the ith row lies in the jth column, all other elements in the jth column are zero.
(d) The first row has a non-zero element.

Example

$$A = \begin{bmatrix} 0 & 1 & -2 & 0 & -1 & 3 \\ 0 & 0 & 0 & 1 & 2 & 5 \\ 0 & 0 & 0 & 0 & 0 & 0 \end{bmatrix}$$

5. *Symmetric matrix.* Here $a_{ij} = a_{ji}$, for all i, j.

Example

$$A = \begin{bmatrix} 2 & -3 & i \\ -3 & 0 & 1 \\ i & 1 & 4 \end{bmatrix}, \quad i^2 = -1$$

6. *Skew-symmetric matrix.* Here $a_{ij} = -a_{ji}$, for all i, j. Obviously $a_{ii} = 0$, for all i.

Example

$$A = \begin{bmatrix} 0 & 1+i & 2 \\ -1-i & 0 & -1 \\ -2 & 1 & 0 \end{bmatrix}, \quad i^2 = -1$$

7. *Hermitian matrix.* Here $a_{ij} = \bar{a}_{ji}$, for all i, j, from which we conclude that the diagonal elements are real.

Example

$$A = \begin{bmatrix} 1 & i & -2 \\ -i & 0 & -1-i \\ -2 & -1+i & 3 \end{bmatrix}, \quad i^2 = -1$$

8. *Skew-Hermitian matrix.* Here $a_{ij} = -\bar{a}_{ji}$, for all i, j, from which we conclude that the diagonal elements are pure imaginary.

Example

$$A = \begin{bmatrix} 0 & 1-i & i \\ -1-i & -i & -3 \\ i & 3 & 2i \end{bmatrix}, \quad i^2 = -1$$

We notice that the last four matrices must be all square in view of their properties.

1.3. Conjugate and transpose of a matrix

The *conjugate* of a matrix A is the matrix obtained by taking the conjugate of all elements of A; and is denoted by $\bar{A}$. Therefore if A is real $\bar{A} = A$, and if A is pure imaginary $\bar{A} = -A$.

Example

$$A = \begin{bmatrix} 1 & i \\ 1+3i & -2 \\ i & 2-i \end{bmatrix}. \quad \text{Then } \bar{A} = \begin{bmatrix} 1 & -i \\ 1-3i & -2 \\ -i & 2+i \end{bmatrix}, \quad i^2 = -1.$$

The *transpose* of a matrix A is the matrix obtained by interchanging the rows of A with the columns of A and vice versa; and is denoted by A^T. So if A is of order (m, n), A^T will be of order (n, m).

Example

$$A = \begin{bmatrix} 1+i & -1 & 3 \\ 0 & 1 & i \end{bmatrix}. \quad \text{Then } A^T = \begin{bmatrix} 1+i & 0 \\ -1 & 1 \\ 3 & i \end{bmatrix}, \quad i^2 = -1.$$

If both operations are simultaneously applied on A we obtain the so-called conjugate transpose of A denoted by A^*.

Example

$$A = \begin{bmatrix} 1 & -i & 3+i \\ -2 & 0 & -4 \end{bmatrix}. \quad \text{Then } A^* = \begin{bmatrix} 1 & -2 \\ i & 0 \\ 3-i & -4 \end{bmatrix}, \quad i^2 = -1.$$

From the above definitions of conjugate and transpose, we can directly see that for a symmetric matrix $A = A^T$, for a skew-symmetric matrix $A = -A^T$, for a Hermitian matrix $A = A^*$ and for a skew-Hermitian matrix $A = -A^*$.

1.4. Operations with matrices

1. *Addition.* If A and B have the same size, i.e. the same number of rows and columns, the element c_{ij} in the matrix

$$C = A + B$$

is given by

$$c_{ij} = a_{ij} + b_{ij}, \quad \text{for all } i, j.$$

Example

$$A = \begin{bmatrix} 1 & -3 \\ 0 & i \\ -1 & 2 \end{bmatrix}, \quad B = \begin{bmatrix} 1 & -1 \\ 0 & 6 \\ 5 & 2 \end{bmatrix}. \quad \text{Then } C = \begin{bmatrix} 2 & -4 \\ 0 & i+6 \\ 4 & 4 \end{bmatrix}, \quad i^2 = -1.$$

2. *Subtraction.* This is similar to addition: if

$$C = A - B,$$

then

$$c_{ij} = a_{ij} - b_{ij}, \quad \text{for all } i, j$$

Example

$$A = \begin{bmatrix} 1 & -2 \\ 0 & 1 \\ 3 & 0 \end{bmatrix}, \quad B = \begin{bmatrix} 1 & 2 \\ 3 & -6 \\ 1 & 7 \end{bmatrix}. \quad \text{Then } C = \begin{bmatrix} 0 & -4 \\ -3 & 7 \\ 2 & -7 \end{bmatrix}$$

3. *Multiplication.* If A is multiplied by a scalar α, the matrix αA is obtained by multiplying all elements of A by α.

Example

$$A = \begin{bmatrix} 1 & -3 \\ 0 & 2 \end{bmatrix}, \quad \alpha = 4. \quad \text{Then } 4A = \begin{bmatrix} 4 & -12 \\ 0 & 8 \end{bmatrix}$$

If A is multiplied by another matrix B, then the element c_{ij} of $C = AB$ is given by

$$c_{ij} = \sum_{k=1}^{n} a_{ik} b_{kj}$$

when n is equal to the number of rows of B or columns of A. Hence the condition for post-multiplying A by B is that the number of columns of A must be equal to the number of rows of B; in that case A is said to be *conformable* to B for multiplication.

Example

$$A = \begin{bmatrix} 1 & 2 & -3 \\ 0 & 1 & 4 \end{bmatrix}, \quad B = \begin{bmatrix} 1 & -2 & 0 & 1 \\ 0 & -3 & 1 & -1 \\ 1 & -1 & 0 & 2 \end{bmatrix}.$$

Then

$$C = \begin{bmatrix} -2 & -5 & 2 & -7 \\ 4 & -7 & 1 & 7 \end{bmatrix}$$

So if A is of order (m, n) and B of order (n, p), then C will be of order (m, p). In general

$$AB \neq BA,$$

even if A is conformable to B and B conformable to A. If

$$AB = BA$$

as a special case, then A and B are said to *commute.*

The above definition for multiplication allows us to raise the power of a matrix to any positive integer q by multiplying A by itself q times. Obviously this can happen only when A is square.

It also enables us to represent the set of linear simultaneous equations

$$\begin{aligned}
a_{11}x_1 + a_{12}x_2 + \cdots + a_{1n}x_n &= b_1 \\
a_{21}x_1 + a_{22}x_2 + \cdots + a_{2n}x_n &= b_2 \\
\vdots \qquad\qquad\qquad & \vdots \\
a_{m1}x_1 + a_{m2}x_2 + \cdots + a_{mn}x_n &= b_m
\end{aligned}$$

in the abbreviated vector-matrix form

$$Ax = b;$$

where

$$A = \begin{bmatrix} a_{11} & a_{12} & \cdots & a_{1n} \\ a_{21} & a_{22} & \cdots & a_{2n} \\ \vdots & & & \vdots \\ a_{m1} & a_{m2} & \cdots & a_{mn} \end{bmatrix}, \quad x = \begin{bmatrix} x_1 \\ x_2 \\ \vdots \\ x_n \end{bmatrix} \quad \text{and } b = \begin{bmatrix} b_1 \\ b_2 \\ \vdots \\ b_m \end{bmatrix}$$

Example

The set of linear equations

$$\begin{aligned}
2x_1 + 3x_2 + 5x_3 &= 1 \\
x_1 - x_2 + 2x_3 &= -1 \\
-x_1 + 2x_2 + x_3 &= 0
\end{aligned}$$

can be represented in the following form

$$\begin{bmatrix} 2 & 3 & 5 \\ 1 & -1 & 2 \\ -1 & 2 & 1 \end{bmatrix} \begin{bmatrix} x_1 \\ x_2 \\ x_3 \end{bmatrix} = \begin{bmatrix} 1 \\ -1 \\ 0 \end{bmatrix}$$

1.5. Partitioning

By partitioning of a matrix A we mean subdivision of it into rectangular blocks of elements called submatrices. For example the matrix $A_{m,n}$ can be partitioned in the following form

$$A = \begin{bmatrix} A_{11} & A_{12} \\ A_{21} & A_{22} \end{bmatrix}$$

where

A_{11} can be of order (l, k)

A_{12} can be of order $(l, n-k)$

A_{21} can be of order $(m-l, k)$

A_{22} can be of order $(m-l, n-k)$

If another matrix $B_{n,p}$ is given by

$$B = \begin{bmatrix} B_{11} & B_{12} \\ B_{21} & B_{22} \end{bmatrix}$$

where

B_{11} is of order (k, q)

B_{12} is of order $(k, p-q)$

B_{21} is of order $(n-k, q)$

B_{22} is of order $(n-k, p-q)$

then

$$AB = \begin{bmatrix} A_{11}B_{11} + A_{12}B_{21} & A_{11}B_{12} + A_{12}B_{22} \\ A_{21}B_{11} + A_{22}B_{21} & A_{21}B_{12} + A_{22}B_{22} \end{bmatrix}$$

Example

$$A = \left[\begin{array}{cc:cc} 1 & -2 & 3 & 0 \\ 1 & -1 & 0 & 2 \\ \hdashline 0 & 3 & 1 & -1 \end{array}\right], \quad B = \left[\begin{array}{cc} 1 & -1 \\ 0 & 2 \\ \hdashline 1 & 0 \\ -2 & -1 \end{array}\right].$$

Then

$$AB = \begin{bmatrix} \begin{bmatrix} 1 & -2 \\ 1 & -1 \end{bmatrix}\begin{bmatrix} 1 & -1 \\ 0 & 2 \end{bmatrix} + \begin{bmatrix} 3 & 0 \\ 0 & 2 \end{bmatrix}\begin{bmatrix} 1 & 0 \\ -2 & -1 \end{bmatrix} \\ [0 \quad 3]\begin{bmatrix} 1 & -1 \\ 0 & 2 \end{bmatrix} + [1 \quad -1]\begin{bmatrix} 1 & 0 \\ -2 & -1 \end{bmatrix} \end{bmatrix} = \begin{bmatrix} 4 & -5 \\ -3 & -5 \\ 3 & 7 \end{bmatrix}$$

If, as a special case, A_{11} and A_{22} are both square and $A_{12} = 0, A_{21} = 0$, then A is called *quasidiagonal* or *block diagonal.*

It turns out from the above, that if A and B are both quasidiagonal, i.e. such that

$$A = \text{quasidiag}\,(A_1, A_2, \ldots, A_m), \quad B = \text{quasidiag}\,(B_1, B_2, \ldots, B_m),$$

and the A_i and B_i matrices are conformable for multiplication, then

$$AB = \text{quasidiag}\,(A_1B_1, A_2B_2, \ldots, A_mB_m)$$

Exercises 1.5

1. Show that $A + B = B + A$.
2. Show $A + (B + C) = (A + B) + C$.
3. Show that $A(BC) = (AB)C$.
4. Show that $A(B + C) = AB + AC$.
5. Show that if $AB = AC$, it does not imply that $B = C$. Think of an example in which $B \neq C$ although $AB = AC$.
6. Show with aid of an example that $AB = 0$ does not imply that $A = 0$ or $B = 0$.
7. Explain why $(A + B)^2 \neq A^2 + B^2 + 2AB$.
8. Show that $A^pA^q = A^{p+q}$ for any positive integers p and q.
9. Show that $(A^q)^p = A^{qp}$ for any positive integers p and q.
10. If $A = \begin{bmatrix} 0 & i \\ i & 0 \end{bmatrix}$, where $i^2 = -1$, give a general expression for A^n.
11. If $AB = -BA$, the matrices A and B are said to be *anticommutative* or to *anticommute*. Show that each of the matrices

$$\begin{bmatrix} 0 & 1 \\ 1 & 0 \end{bmatrix}, \begin{bmatrix} 0 & -i \\ i & 0 \end{bmatrix}, \begin{bmatrix} 1 & 0 \\ 0 & -1 \end{bmatrix} \quad (i^2 = -1)$$

anticommutes with the others.

12. Show that if $A = \begin{bmatrix} \cos\theta & \sin\theta \\ -\sin\theta & \cos\theta \end{bmatrix}$, then $A^n = \begin{bmatrix} \cos n\theta & \sin n\theta \\ -\sin n\theta & \cos n\theta \end{bmatrix}$.
13. Show that $A + 0 = A$ and that $AI = A$.
14. Show that any diagonal matrices of the same order commute.
15. A matrix A for which $A^2 = I$ is called *involutory*. Show that A is involutory if and only if $(I - A)(I + A) = 0$.
16. Show that $(\overline{\overline{A}}) = A$ and that $(A^T)^T = A$.
17. Show that $(\bar{A})^T = (\overline{A^T}) = A^*$.
18. Show that $(\overline{AB}) = \bar{A}\bar{B}$.
19. Show that $(A + B)^T = A^T + B^T$.
20. Show that $(AB)^T = B^TA^T$.
21. Show that $(\bar{A})^n = (\overline{A^n})$.
22. Show that $(A^n)^T = (A^T)^n$.
23. Show that $A^pB^q = B^qA^p$ if $AB = BA$.
24. Show that, if $A^*A = 0$, then $A = 0$.
25. Show that, if $Ax = 0$ for all vectors x, then $A = 0$.
26. Show that $(A_1A_2 \ldots A_p)^T = A_p^TA_{p-1}^T \ldots A_1^T$.
27. If $A_{n,n} = \begin{bmatrix} 0 & 1 & & & \\ & 0 & 1 & & \\ & & \ddots & \ddots & \\ & & & \ddots & 1 \\ & & & & 0 \end{bmatrix}$, find A^p with $p < n$.
28. Show that A^TA is symmetric and that A^*A is Hermitian.
29. If A is Hermitian, show that P^*AP is also Hermitian.

30. If A and B are real and $A + ib$ is Hermitian with $i^2 = -1$, find properties of A and B.
31. Show that $A + A^T$ is symmetric and that $A + A^*$ is Hermitian.
32. Show that $A - A^T$ is skew-symmetric and that $A - A^*$ is skew-Hermitian.
33. If A and B are symmetric, find the conditions for AB to be (a) symmetric and (b) skew-symmetric.
34. If A and B are symmetric matrices, show that $AB + BA$ is symmetric and that $AB - BA$ is skew-symmetric.
35. Show that any square matrix A can be written as $A = A^s + A^{ss}$, where A^s and A^{ss} are respectively the symmetric and the skew-symmetric parts of A. Find A^s and A^{ss}.
36. Show that the Hermitian part A^H and the skew-Hermitian part A^{sH} of a complex square matrix A are given by $A^H = \dfrac{A + A^*}{2}$ and $A^{sH} = \dfrac{A - A^*}{2}$.
37. A square matrix A is called an *upper triangular matrix* if $a_{ij} = 0$ for all $j < i$. If $a_{ij} = 0$ for all $j > i$, A is called a *lower triangular matrix*. Show that if A is an upper triangular matrix, then A^T is a lower triangular matrix and vice-versa.
38. A square matrix A is called an *upper matrix* if $a_{ij} = 0$ for all $j \leqslant i$. If $a_{ij} = 0$ for $j \geqslant i$, A is called a *lower matrix*. Show that for both types of matrix $A^p = 0$, if p is large enough.
39. If the transfer matrices for the two-port networks N_1 and N_2 are given by

$$\begin{bmatrix} v_1 \\ i_1 \end{bmatrix} = \begin{bmatrix} a & b \\ c & d \end{bmatrix} \begin{bmatrix} v_2 \\ -i_2 \end{bmatrix}, \quad \begin{bmatrix} v_1' \\ i_1' \end{bmatrix} = \begin{bmatrix} a' & b' \\ c' & d' \end{bmatrix} \begin{bmatrix} v_2' \\ -i_2' \end{bmatrix}$$

find the transfer matrix of the cascaded network N_1N_2.

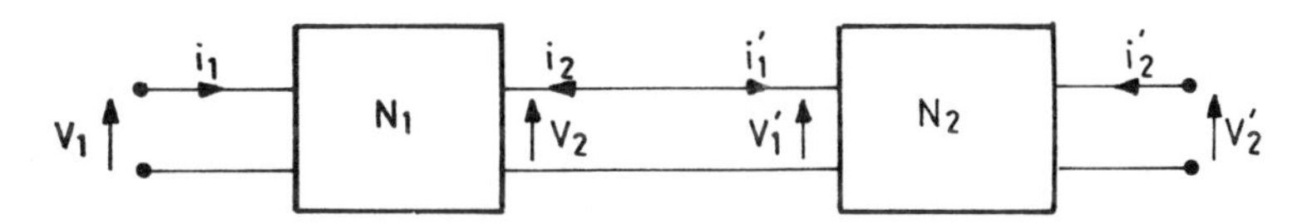

Obtain also the input impedance of the cascade $z_{in} = v_1/i_1$, if a load R is connected at the output and is given by $v_2' = -Ri_2'$.
40. The difference equation $x(k) = Ax(k-1) + y(k)$ appears in the formulation of many socio-economic systems, where the state vector x is only measured at some discrete points if k is an integer representing years, weeks. . . . Show that the solution of the equation at some time n is given by

$$x(n) = A^n x(0) + \sum_{j=1}^{n} A^{n-j} y(j).$$

In examples of demography or education, x represents respectively the population divided into age categories or the number of students enrolled at different levels of education; whereas A is the state matrix whose elements depend on fertility, mortality and migration rates in demographic studies or promotion, repeater and drop-out ratios in education systems. Show, neglecting drop-out ratios, that for an education system of m levels whose difference equation for the ith level is $E_i(k) = r_i E_i(k-1) + p_{i-1} E_{i-1}(k-1)$, the bidiagonal matrix A and vector y are given by

$$A = \begin{bmatrix} r_1 & & & \\ p_1 & r_2 & & \\ & \ddots & \ddots & \\ & & p_{m-1} & r_m \end{bmatrix}, \quad y(k) = \begin{bmatrix} a(k) \\ 0 \\ \vdots \\ 0 \end{bmatrix}$$

And if no students are admitted at the first level after a year t, find the number of years that should be allowed after that, in order to deplete the system from students; assuming that no students fail ($r = 0$) during these years. Check your result using the matrix A. For further reading about difference equations and relevant applications to education and population refer to Thonstad (1969) and Pollard (1973), and for application to manpower planning via education policy control, see Deif (1977, 1980).

CHAPTER TWO

SCALAR FUNCTIONS OF MATRICES

Up till now we have been handling matrices in their form of rectangular arrays. However we can deduce from them some scalar functions which are extremely useful in algebra and analysis. In the coming sections we discuss the most important ones.

2.1. Determinant of a matrix

The *determinant* of a square matrix A is a scalar quantity denoted by 'det A' or '$|A|$' and is given by

$$|A| = \sum_{j_1,j_2,\ldots,j_n} p(j_1,j_2,\ldots j_n)\, a_{1j_1}a_{2j_2}\ldots a_{nj_n}$$

where $p(j_1, j_2, \ldots, j_n)$ is a permutation equal to ± 1. For an (n, n) matrix there exist $n!$ permutations. For a matrix of order $(3, 3)$

$$p(1, 2, 3) = 1, \quad p(1, 3, 2) = -1, \quad p(3, 1, 2) = 1$$
$$p(3, 2, 1) = -1, \quad p(2, 3, 1) = 1, \quad p(2, 1, 3) = -1.$$

In general

$$p(j_1, j_2, \ldots, j_n) = \text{sign} \prod_{1 \leqslant s < r \leqslant n} (j_r - j_s)$$

Example

$$\begin{vmatrix} a_{11} & a_{12} & a_{13} \\ a_{21} & a_{22} & a_{23} \\ a_{31} & a_{32} & a_{33} \end{vmatrix} = \begin{aligned} &a_{11}a_{22}a_{33} - a_{11}a_{23}a_{32} + a_{13}a_{21}a_{32} \\ &-a_{13}a_{22}a_{31} + a_{12}a_{23}a_{31} - a_{12}a_{21}a_{33} \end{aligned}$$

The determinant can also be written in a much simpler form but still relying on the definition of a determinant like the following

$$|A| = \sum_{j=1}^{n} a_{ij}C_{ij}, \quad i = \text{constant}$$

$$\text{or} = \sum_{i=1}^{n} a_{ij}\,C_{ij}, \quad j = \text{constant}$$

where C_{ij} is called the *cofactor* of the element a_{ij} and is given by

$$C_{ij} = \sum_{j_1, j_2, \ldots, j_n} p(j_1, j_2, \ldots, j_n) \prod_{\substack{k=1 \\ k \neq i}}^{n} a_{kj_k}$$

C_{ij} is also equal to $(-)^{i+j} M_{ij}$, where M_{ij} is called the *minor* and is equal to the determinant of the submatrix obtained from the matrix A by deleting the ith row and jth column. For the example cited above,

$$\begin{aligned} C_{12} &= a_{23}a_{31} - a_{21}a_{33} \\ &= -M_{12} \end{aligned}$$

When the rows and columns struck out of A have the same indices, the resulting submatrix is located symmetrically with respect to the main diagonal of A; and we call the corresponding minor a *principal* minor of A. For the example cited above M_{22} is a principal minor, whereas the minor

$$\begin{vmatrix} a_{11} & \cdots & a_{1r} \\ \vdots & & \vdots \\ a_{r1} & & a_{rr} \end{vmatrix}$$

is called a *leading* principal minor of A of order r.

It is customary to call a square matrix A a *nonsingular* matrix if $|A| \neq 0$, whereas if $|A| = 0$ the matrix A is called a *singular* matrix.

Properties of determinants

A survey of these properties can be found in Franklin (1968), p. 5. We list them in the following order.

1. $|A| = |A^T|$

Proof

$$|A| = \sum_{j_1, j_2, \ldots, j_n} p(j_1, j_2, \ldots, j_n)\, a_{1j_1}a_{2j_2} \cdots a_{nj_n}$$

$$|A^T| = \sum_{j_1, j_2, \ldots, j_n} p(j_1, j_2, \ldots, j_n)\, a_{j_1 1}a_{j_2 2} \cdots a_{j_n n}$$

However

$$a_{j_1 1} \cdots a_{j_n n} = a_{1i_1} \cdots a_{ni_n}$$

by rearranging the elements; for there exists a one-to-one correspondence between permutations j and permutations i. Also

$$p(j_1, j_2, \ldots, j_n) = p(i_1, i_2, \ldots, i_n).$$

Therefore

$$|A^T| = \sum_{i_1, i_2, \ldots, i_n} p(i_1, \ldots, i_n)\, a_{1i_1} a_{2i_2} \ldots a_{ni_n} = |A|$$

Example

$$\begin{vmatrix} 1 & 0 & -1 \\ 2 & 1 & 0 \\ 1 & -1 & 2 \end{vmatrix} = \begin{vmatrix} 1 & 2 & 1 \\ 0 & 1 & -1 \\ -1 & 0 & 2 \end{vmatrix} = 5$$

2. If any vector in A is multiplied by a scalar α, the determinant of the new matrix is equal to $\alpha|A|$. The proof follows directly by noticing that $a_{1j_1} \ldots a_{nj_n}$ must contain one element belonging to the vector.

Example

$$\begin{vmatrix} 1 & -1 \\ 2 & 3 \end{vmatrix} = 5, \quad \text{Then} \quad \begin{vmatrix} 4 & -4 \\ 2 & 3 \end{vmatrix} = 20$$

3. If any vector in A is a zero vector then $|A| = 0$. The proof follows from property 2, by taking $\alpha = 0$.

Example

$$\begin{vmatrix} 0 & 0 & 0 \\ 1 & -1 & 2 \\ 0 & 1 & 3 \end{vmatrix} = 0$$

4. If two rows in A or if two columns in A are interchanged, the sign of the determinant is reversed.

Proof. The proof will be accomplished on a row basis but is similar on a column basis. Let the two rows r and s be interchanged in A to produce another matrix B. Then

$$b_{rj} = a_{sj}, b_{sj} = a_{rj} \quad \text{and} \quad b_{ij} = a_{ij} \quad \text{for all } i \neq r \text{ or } s.$$

Now

$$|B| = \sum_{j_1, j_2, \ldots, j_n} p(j_1, \ldots, j_r, \ldots, j_s, \ldots, j_n)\, b_{1j_1} \ldots b_{rj_r} \ldots b_{sj_s} \ldots b_{nj_n}$$

$$|A| = \sum_{j_1, j_2, \ldots, j_n} p(j_1, \ldots, j_s, \ldots, j_r, \ldots, j_n)\, a_{1j_1} \ldots a_{rj_r} \ldots a_{sj_s} \ldots a_{nj_n}$$

But

$$p(j_1, \ldots, j_r, \ldots, j_s, \ldots, j_n) = -p(j_1, \ldots, j_s, \ldots, j_r, \ldots, j_n)$$

and

$$\begin{aligned} b_{1j_1} \ldots b_{rj_r} \ldots b_{sj_s} \ldots b_{n,j_n} &= a_{1j_1} \ldots a_{sj_s} \ldots a_{rj_r} \ldots a_{nj_n} \\ &= a_{1j_1} \ldots a_{rj_r} \ldots a_{sj_s} \ldots a_{nj_n} \end{aligned}$$

Hence

$$|B| = -|A|$$

Example

$$\begin{vmatrix} 1 & 0 & -1 \\ 2 & 1 & 1 \\ 0 & 1 & 2 \end{vmatrix} = -1. \quad \text{Then} \quad \begin{vmatrix} 0 & 1 & -1 \\ 1 & 2 & 1 \\ 1 & 0 & 2 \end{vmatrix} = 1.$$

5. If two rows or columns in A are identical then $|A| = 0$. The proof follows from property 4, by interchanging the two identical vectors.

Example

$$\begin{vmatrix} 1 & 5 & -1 \\ 0 & 1 & 3 \\ 1 & 5 & -1 \end{vmatrix} = 0.$$

6. Let the element $a_{sj} = c_{sj} + b_{sj}$, for all $j = 1, \ldots, n$. Then the determinant of A is equal to the sum of two determinants; the first is obtained by replacing a_{sj} in A by c_{sj} for all $j = 1, \ldots, n$ and the second is obtained by replacing a_{sj} in A by b_{sj} for all $j = 1, \ldots, n$.

Proof

$$|A| = \sum_{j_1,\ldots,j_n} p(j_1, \ldots, j_n)\, a_{1j_1} \ldots a_{nj_n}$$

$$= \sum_{j_1,\ldots,j_n} p(j_1, \ldots, j_n)\, a_{1j_1} \ldots (c_{sj_s} + b_{sj_s}) \ldots a_{nj_n}$$

$$= \sum_{j_1,\ldots,j_n} p(j_1, \ldots, j_n)\, a_{1j_1} \ldots c_{sj_s} \ldots a_{nj_n} +$$

$$+ \sum_{j_1,\ldots,j_n} p(j_1, \ldots, j_n)\, a_{1j_1} \ldots b_{sj_s} \ldots a_{nj_n}$$

Example

$$\begin{vmatrix} 1 & 2 & -1 \\ 7 & 9 & 13 \\ 5 & -1 & 3 \end{vmatrix} = \begin{vmatrix} 1 & 2 & -1 \\ 8 & 7 & 12 \\ 5 & -1 & 3 \end{vmatrix} + \begin{vmatrix} 1 & 2 & -1 \\ -1 & 2 & 1 \\ 5 & -1 & 3 \end{vmatrix}$$

7. $|AB| = |A|\,|B|$

Proof. Let $C = AB$; then

$$|C| = \sum_{j_1,\ldots,j_n} p(j_1, \ldots, j_n)\, c_{1j_1} \ldots c_{nj_n}$$

but

$$c_{1j_1} = \sum_{k_1=1}^{n} a_{1k_1} b_{k_1 j_1}, \quad c_{2j_2} = \sum_{k_2=1}^{n} a_{2k_2} b_{k_2 j_2}, \ldots$$

Therefore

$$|C| = \sum_{k_1=1}^{n} \ldots \sum_{k_n=1}^{n} a_{1k_1} \ldots a_{nk_n} \sum_{j_1,\ldots,j_n} p(j_1, \ldots, j_n)\, b_{k_1 j_1} \ldots b_{k_n j_n}$$

but the last sum is equal to a determinant not equal to zero if $k_1, \ldots, k_n$ form a permutation k of the numbers $1, 2, \ldots, n$. Therefore in the above summation we can eliminate all of the n^n combinations $k_1, \ldots, k_n$ except the $n!$ permutations k. Therefore

$$|C| = \sum_{k_1,\ldots,k_n} a_{1k_1} \ldots a_{nk_n} \sum_{j_1,\ldots,j_n} p(j_1, \ldots, j_n)\, b_{k_1 j_1} \ldots b_{k_n j_n}.$$

Now if

$$\sum_{j_1,\ldots,j_n} p(j_1, \ldots, j_n)\, b_{k_1 j_1} \ldots b_{k_n j_n}$$

is written as

$$\sum_{j_1,\ldots,j_n} p(k_1, \ldots, k_n)\, p(j_1, \ldots, j_n)\, b_{1j_1} \ldots b_{nj_n},$$

then we obtain

$$|C| = \sum_{k_1,\ldots,k_n} p(k_1, \ldots, k_n)\, a_{1k_1} \ldots a_{nk_n} \cdot |B| = |A|\,|B|.$$

Example

$$A = \begin{bmatrix} 1 & 0 & -1 \\ 1 & 2 & 2 \\ 0 & 1 & 0 \end{bmatrix}, \quad B = \begin{bmatrix} 2 & 0 & 0 \\ -1 & 1 & 1 \\ 0 & 1 & 2 \end{bmatrix}$$

Then

$$AB = \begin{bmatrix} 2 & -1 & -2 \\ 0 & 4 & 6 \\ -1 & 1 & 1 \end{bmatrix}$$

$$|A| = -3, \quad |B| = 2, \quad |AB| = -6.$$

8. $\sum_{j=1}^{n} a_{ij} C_{kj} = 0$, for all $i \neq k$, i.e. the sum of the product of the elements of a row by the cofactors of another row is equal to zero.

Proof. $\sum_{j=1}^{n} a_{ij} C_{kj}$ is equal to the determinant of the matrix obtained by replacing the elements of the kth row by those of the ith row. But this equal to zero due to the equality of the elements of the two rows.

Example

$$A = \begin{bmatrix} 1 & 2 & -1 \\ 0 & 1 & -2 \\ -3 & 1 & 0 \end{bmatrix}$$

$$\sum_{j=1}^{n} a_{1j} C_{2j} = 1(-1) + 2(-3) - 1(-7) = 0.$$

9. The determinant of a matrix A does not change if the elements of a certain row are multiplied by a constant and then added to any other row.

Proof. The determinant of the new matrix A' is given by

$$\begin{aligned} |A'| &= \sum_j (a_{ij} + \alpha a_{kj}) C_{ij}, \quad i \neq k \\ &= \sum_j a_{ij} C_{ij} + \alpha \sum_j a_{kj} C_{ij} \\ &= \sum_j a_{ij} C_{ij} = |A| \end{aligned}$$

Example

$$\begin{vmatrix} 1 & 2 & -1 \\ 0 & 1 & -2 \\ -3 & 1 & 0 \end{vmatrix} = \begin{vmatrix} 1 & 4 & -5 \\ 0 & 1 & -2 \\ -3 & 1 & 0 \end{vmatrix}$$

10. $(d/d\lambda)|A(\lambda)|$ = sum of the determinants where each of them is obtained by differentiating the rows of A w.r.t. λ one at a time, then taking its determinant.

Proof

$$|A(\lambda)| = \sum_{j_1, \ldots, j_n} p(j_1, \ldots, j_n)\, a_{1j_1}(\lambda) \ldots a_{nj_n}(\lambda).$$

Hence

$$\begin{aligned} \frac{d}{d\lambda}|A| = &\sum_{j_1, \ldots, j_n} p(j_1, \ldots, j_n) \frac{da_{1j_1}(\lambda)}{d\lambda} a_{2j_2}(\lambda) \ldots a_{nj_n}(\lambda) + \\ &+ \sum_{j_1, \ldots, j_n} p(j_1, \ldots, j_n)\, a_{1j_1}(\lambda) \frac{da_{2j_2}(\lambda)}{d\lambda} \ldots a_{nj_n}(\lambda) + \cdots + \\ &+ \sum_{j_1, \ldots, j_n} p(j_1, \ldots, j_n)\, a_{1j_1}(\lambda) \ldots a_{n-1 j_{n-1}}(\lambda) \frac{da_{nj_n}}{d\lambda}(\lambda) \end{aligned}$$

Example

$$\frac{d}{dt}\begin{vmatrix} e^t & \cos t \\ 1 & \sin\sqrt{t} \end{vmatrix} = \begin{vmatrix} e^t & -\sin t \\ 1 & \sin\sqrt{t} \end{vmatrix} + \begin{vmatrix} e^t & \cos t \\ 0 & \cos\sqrt{t}/2\sqrt{t} \end{vmatrix}.$$

The reader can notice that properties 8, 9, 10 are also valid on a column basis.

Exercises 2.1

1. Calculate the determinants of the following matrices

$$\begin{bmatrix} 1 & 2 & -1 \\ 0 & 1 & -1 \\ 1 & 2 & -4 \end{bmatrix}, \quad \begin{bmatrix} 1 & 0 & 1 & -1 \\ 1 & 0 & 1 & 2 \\ -1 & 3 & 1 & 0 \\ -1 & 3 & 5 & -7 \end{bmatrix}, \quad \begin{bmatrix} a_{11} & & & \\ a_{21} & a_{22} & & 0 \\ a_{31} & a_{32} & \ddots & \\ \vdots & & & \\ a_{n1} & \dots & & a_{nn} \end{bmatrix}.$$

2. Differentiate w.r.t. λ

$$\begin{vmatrix} e^{\lambda} & 1 & \cos\lambda \\ \cos\lambda & \lambda & \sin\lambda \\ e^{-\lambda} & 1 & 0 \end{vmatrix}$$

3. Show that $|\alpha A| = \alpha^n |A|$, where n is the order of A and α is a scalar.
4. Show that $|A^n| = |A|^n$.
5. A matrix A is called *idempotent* if $A^k = A$, for some positive integer k; find $|A|$.
6. A matrix A is called *nilpotent* if $A^k = 0$, for some positive integer k; find $|A|$.
7. If $A^n = I$, find $|A|$.
8. If $|ABC| = |A| \neq 0$, find $|BC|$.
9. Show that $|\bar{A}| = \overline{|A|}$.
10. Show that $|AB| = |A^T B| = |A^T B^T|$, where A and B are square.
11. If A is skew-symmetric and of odd order, show that $|A| = 0$.
12. If AB is nonsingular, with A and B being square, show that A and B are both nonsingular matrices.
13. If there exists a matrix A^{-1} such that $A^{-1}A = AA^{-1} = I$, show that A^{-1} is nonsingular if A is nonsingular; i.e. if $|A| \neq 0$, there exists a matrix A^{-1} such that $A^{-1}A = AA^{-1} = I$. Show also that $|A^{-1}| = 1/|A|$.
14. *Prove the following important theorem*: if $|A| \neq 0$, then $Ax = 0$ implies $x = 0$. *Hint*: write equation $Ax = 0$ in the form $\Sigma_{j=1}^n a_{1j}x_j = 0$, $\Sigma_{j=1}^n a_{2j}x_j = 0, \dots$ Multiply the equations respectively by $C_{11}, C_{21}, \dots$, then add them; using property 8 of determinants, we find that $x_1 = 0$. To prove that $x_2 = 0$, multiply the equations by $C_{12}, C_{22}, \dots$, etc.
15. Show that

$$\begin{vmatrix} \lambda_1^{n-1} & \lambda_1^{n-2} & \dots & \lambda_1 & 1 \\ \lambda_2^{n-1} & \lambda_2^{n-2} & \dots & \lambda_2 & 1 \\ \vdots & \vdots & & \vdots & \vdots \\ \lambda_n^{n-1} & \lambda_n^{n-2} & \dots & \lambda_n & 1 \end{vmatrix} = \prod_{1 \leqslant i < j \leqslant n} (\lambda_i - \lambda_j)$$

This matrix is called a Vandermonde matrix.

16. Show that the determinant of a Hermitian matrix is real and that of a skew-Hermitian matrix is imaginary.

17. If u and v are two vectors of dimension n, show that uv^T is singular.

18. If A and B are respectively of order (m, n) and (n, m), where $m > n$, show that AB is singular.

19. A *major* determinant of a matrix of order (p, q) is any determinant of maximum order of the matrix. For example

$$\begin{vmatrix} a_{11} & a_{13} \\ a_{21} & a_{23} \end{vmatrix} \text{ is a major determinant of } \begin{bmatrix} a_{11} & a_{12} & a_{13} \\ a_{21} & a_{22} & a_{23} \end{bmatrix}$$

in fact, the determinant of a square matrix is its major determinant. Suppose that A is of order (m, n) and B of order (n, m), where $m \leqslant n$. Then a major determinant of A and a major determinant of B are corresponding majors if and only if the columns of A used to form the major of A have the same indices as do the rows of B used to form the major of B. Now prove the following theorem: 'If A is of order (m, n) and B of order (n, m) and $m \leqslant n$, show that $|AB|$ is equal to the sum of the products of the corresponding majors of A and B'. To clarify the theorem to the reader we give the following example:

$$\begin{bmatrix} x_1 & x_2 & x_3 \\ y_1 & y_2 & y_3 \end{bmatrix} \begin{bmatrix} u_1 & v_1 \\ u_2 & v_2 \\ u_3 & v_3 \end{bmatrix} = \begin{vmatrix} x_1 & x_2 \\ y_1 & y_2 \end{vmatrix} \begin{vmatrix} u_1 & v_1 \\ u_2 & v_2 \end{vmatrix} + \begin{vmatrix} x_1 & x_3 \\ y_1 & y_3 \end{vmatrix} \begin{vmatrix} u_1 & v_1 \\ u_3 & v_3 \end{vmatrix} + \begin{vmatrix} x_2 & x_3 \\ y_2 & y_3 \end{vmatrix} \begin{vmatrix} u_2 & v_2 \\ u_3 & v_3 \end{vmatrix}$$

This expansion is due to Binet–Cauchy.

20. If $i_1 i_2 \ldots i_r$, $n > r \geqslant 1$, is any set of r numbers chosen from $1,2, \ldots, n$ and arranged in natural order, and if $k_1 k_2 \ldots k_{n-r}$ is the set of $n - r$ remaining integers also arranged in natural order, then $i_1, i_2 \ldots i_r$ and $k_1 k_2 \ldots k_{n-r}$ are called *complementary* sets of indices. For example, if $n = 8$, then 147 and 23568 are such complementary sets of indices. If $i_1 i_2 \ldots i_r$ and $k_1 k_2 \ldots k_{n-r}$ are complementary sets of row indices and if $j_1 j_2 \ldots j_r$ and $p_1\, p_2 \ldots p_{n-r}$ are complementary sets of column indices, the two minor determinants $M_{i_1 i_2 \ldots i_r, j_1 j_2 \ldots j_r}$ and $M_{k_1 k_2 \ldots k_{n-r}, p_1, p_2 \ldots p_{n-r}}$, or as we shall write them $M_{(i)(j)}$ and $M_{(k)(p)}$, are called complementary minors of A. For example if $n = 5$, we have as complementary minors:

$$M_{23,24} = \begin{vmatrix} a_{22} & a_{24} \\ a_{32} & a_{34} \end{vmatrix} \quad \text{and } M_{145,135} = \begin{vmatrix} a_{11} & a_{13} & a_{15} \\ a_{41} & a_{43} & a_{45} \\ a_{51} & a_{53} & a_{55} \end{vmatrix}.$$

Notice that if we strike out of A the complete rows and columns of A represented in one minor, there remain just the elements of the complementary minor. Now we define the algebraic complement of $M_{(i)(j)}$ as $A_{(i)(j)}$, and it is given by

$$A_{(i)(j)} = (-1)^{\Sigma i + \Sigma j} M_{(i)(j)}$$

Now prove Laplace's expansion of $|A|$ as follows: 'If $i_1 i_2 \ldots i_r$ is any fixed set of r row indices in the natural order, and if $j_1 j_2 \ldots j_r$ runs over every set of r column indices, each set being also in the natural order, we have

$$|A| = \sum_{(j)} M_{(k)(p)} A_{(i)(j)}$$

where the is and ks, and js and ps are complementary sets of row and column indices respectively, and where the summation extends over all the sets of r column indices described above'. To clarify the theorem by an example:

$$\begin{vmatrix} 1 & 0 & 2 & 0 \\ 2 & 0 & 1 & 0 \\ 0 & 4 & 0 & 3 \\ 0 & 3 & 0 & 4 \end{vmatrix} = \begin{vmatrix} 1 & 0 \\ 2 & 0 \end{vmatrix} \cdot (-1)^{1+2+1+2} \begin{vmatrix} 0 & 3 \\ 0 & 4 \end{vmatrix} + \begin{vmatrix} 1 & 2 \\ 2 & 1 \end{vmatrix}$$
$$\cdot (-)^{1+2+1+3} \begin{vmatrix} 4 & 3 \\ 3 & 4 \end{vmatrix} + \begin{vmatrix} 1 & 0 \\ 2 & 0 \end{vmatrix}$$
$$\cdot (-1)^{1+2+1+4} \begin{vmatrix} 4 & 0 \\ 3 & 0 \end{vmatrix} + \begin{vmatrix} 0 & 2 \\ 0 & 1 \end{vmatrix}$$
$$\cdot (-1)^{1+2+2+3} \begin{vmatrix} 0 & 3 \\ 0 & 4 \end{vmatrix} + \begin{vmatrix} 0 & 0 \\ 0 & 0 \end{vmatrix}$$
$$\cdot (-1)^{1+2+2+4} \begin{vmatrix} 0 & 0 \\ 0 & 0 \end{vmatrix} + \begin{vmatrix} 2 & 0 \\ 1 & 0 \end{vmatrix}$$
$$\cdot (-1)^{1+2+3+4} \begin{vmatrix} 0 & 4 \\ 0 & 3 \end{vmatrix} = 21$$

21. Show that $\begin{vmatrix} A & 0 \\ 0 & B \end{vmatrix} = |A|\,|B|$, if A and B are square.

22. Show that $\begin{vmatrix} A & 0 \\ C & B \end{vmatrix} = |A|\,|B|$, if A and B are square.

23. Let M be a $(2n, 2n)$ matrix partitioned into (n, n) matrices:

$$M = \begin{bmatrix} A & B \\ C & D \end{bmatrix}$$

where $|D| \neq 0$, and D and C commute. Show that $|M| = |AD - BC|$.
Hint: post-multiply by $\begin{bmatrix} I & 0 \\ L & I \end{bmatrix}$, where L is suitably chosen.

24. Using the relation $\begin{bmatrix} A & B \\ C & D \end{bmatrix} \begin{bmatrix} D & -B \\ -C & A \end{bmatrix} = \begin{bmatrix} AD - BC & BA - AB \\ CD - DC & DA - CB \end{bmatrix}$,

find under what condition does $\begin{vmatrix} A & B \\ C & D \end{vmatrix} = |AD - BC|$? A, B, C, and D need not be nonsingular.

2.2. Trace of a matrix

The *trace* of a square matrix A is a scalar quantity obtained by summing up the diagonal elements of A, and is denoted by 'tr A'. i.e.

$$\operatorname{tr} A = \sum_{i=1}^{n} a_{ii}$$

Example

$$A = \begin{bmatrix} 1 & 2 & 3 \\ -1 & -5 & 0 \\ 6 & -2 & -4 \end{bmatrix}. \quad \text{Then } \operatorname{tr} A = 1 + (-5) + (-4) = -8.$$

Properties of the trace of a matrix are given as exercises for the reader.

Exercises 2.2

1. Show that $\operatorname{tr} A = \operatorname{tr} A^T$.
2. Show that $\operatorname{tr}(A + B) = \operatorname{tr} A + \operatorname{tr} B$.
3. Explain why $\operatorname{tr}(AB) \neq \operatorname{tr} A \times \operatorname{tr} B$?
4. Is $\operatorname{tr}(A)^n = (\operatorname{tr} A)^n$ true?
5. Show that $\operatorname{tr}(\alpha A) = \alpha \operatorname{tr} A$, with α a scalar.
6. If $A = \begin{bmatrix} \cos\theta & \sin\theta \\ \sin\theta & -\cos\theta \end{bmatrix}$, find $\operatorname{tr} A^n$.
7. If $\operatorname{tr} A^*A = 0$, show that $A = 0$.
8. Show that, if $\operatorname{tr}(AB) = 0$ for all matrices B, then $A = 0$.
9. Show that $\operatorname{tr} AB = \operatorname{tr} BA$, for any two matrices A and B, as long as the products exist.
10. Show that if $BB^{-1} = I$, then $\operatorname{tr}(B^{-1}AB) = \operatorname{tr} A$.
11. Show that, for any m matrices $A_1, A_2, \ldots, A_m$ of order n, the trace of $A_1A_2 \ldots A_m$ remains unchanged after a cyclic permutation of the factors $A_1, A_2, \ldots, A_m$.
12. Show that $\operatorname{tr}(AB)^n = \operatorname{tr}(BA)^n$, when n is a positive integer.
13. Show that $|I + \alpha A| = I + \alpha \operatorname{tr} A + \ldots = \exp\left(\sum_{k=1}^{\infty} \frac{(-)^{k-1}}{k} \alpha^k \operatorname{tr} A^k\right)$.
14. If A is of order (m, n) and B of order (n, m), show that $|I + AB| = |I + BA|$. *Hint*: take $\alpha = 1$ in Exercise 13.

2.3. Vector norms and related matrix norms

The notion of vector norm is introduced as a scalar function to consider the magnitude or length of a vector in $\mathbb{C}^n$; exactly like we associate the number $|\alpha|$ with each complex number α.

The *norm* of a vector $x \in \mathbb{C}^n$, denoted by $\|x\|$, is a non-negative real scalar which satisfies the following relations:

1. $\|x\| > 0$, if $x \neq 0$ (positivity)
2. $\|\alpha x\| = |\alpha| \, \|x\|$, for any $\alpha \in \mathbb{C}$ and any $x \in \mathbb{C}^n$ (homogeneity)
3. $\|x + y\| \leqslant \|x\| + \|y\|$, for all vectors $x, y \in \mathbb{C}^n$ (triangular inequality)

The most frequently used vector norm is the Holder norm and is given by

$$\| x \|_p = \left(\sum_{i=1}^{n} |x_i|^p \right)^{1/p},$$

where p is a positive integer.

The most popular and widely used values of the Holder norm are

(a) $\| x \|_1 = \sum_{i=1}^{n} |x_i|$

Example

$$x = \begin{bmatrix} 1+i \\ -1 \\ 3 \end{bmatrix}. \quad \text{Then } \| x \|_1 = \sqrt{2} + 1 + 3 = 4 + \sqrt{2}.$$

(b) $\| x \|_2 = \sqrt{\sum_{i=1}^{n} |x_i|^2} = \sqrt{x^*x}$, usually referred to as the Euclidian norm.

Example

$$x = \begin{bmatrix} i \\ i-1 \\ 1 \end{bmatrix}. \quad \text{Then } \| x \|_2 = \sqrt{1+2+1} = 2.$$

The reader can see that the Euclidian norm of a vector is equal to its length in $\mathbb{C}^n$. It is a matter of interest too to determine the linear transformation $x = Uy$ which leaves the Euclidian norm unchanged. In other words can we find a matrix U such that

$$x^*x = (Uy)^*(Uy) = y^*y?$$

The answer is in the affirmative if

$$U^*U = I.$$

A matrix U which satisfies the above relation is called a *unitary matrix.* If U is real then it is called an *orthogonal matrix.* The reader can prove that U^* is unitary whenever U is, that the determinant of U has an absolute value 1 and that the product of unitary matrices is again unitary.

(c) $\| x \|_\infty = \max_i |x_i|$

Example

$$x = \begin{bmatrix} 1-3i \\ 4 \\ -2 \end{bmatrix}. \quad \text{Then } \| x \|_\infty = 4.$$

To show that the Holder norm $\| x \|_p$ satisfies the above three relations is not difficult. To satisfy the first two is easy, and so we direct ourselves to prove the third.

Proof

$$(\| x + y \|_p)^p = \sum_{i=1}^{n} |x_i + y_i|^p = \sum_{i=1}^{n} |(x_i + y_i)^p|$$

$$= \sum_{i=1}^{n} |x_i^p + (p_1)\, x_i^{p-1} y_i + \cdots + y_i^p|$$

$$\leqslant \sum_{i=1}^{n} |x_i|^p + (p_1) \sum_{i=1}^{n} |x_i|^{p-1} |y_i| + \cdots + \sum_{i=1}^{n} |y_i|^p$$

Using the Holder inequality explained in Appendix 1, we obtain

$$(\| x + y \|_p)^p \leqslant \sum_{i=1}^{n} |x_i|^p + (p_1) \left(\sum_{i=1}^{n} |x_i|^p \right)^{(p-1)p} \left(\sum_{i=1}^{n} |y_i|^p \right)^{1/p} + \cdots + \sum_{i=1}^{n} |y_i|^p$$

$$= \left(\left(\sum_{i=1}^{n} |x_i|^p \right)^{1/p} + \left(\sum_{i=1}^{n} |y_i|^p \right)^{1/p} \right)^p$$

Hence

$$\| x + y \|_p \leqslant \| x \|_p + \| y \|_p$$

The norm of a square matrix A, denoted by $\| A \|$, is a non-negative real scalar which satisfies the following relations:

1. $\| A \| > 0$, if $A \neq 0$.
2. $\| \alpha A \| = |\alpha| \| A \|$, for any $\alpha \in \mathbb{C}$ and any $A \in \mathbb{C}^{nn}$
3. $\| A + B \| \leqslant \| A \| + \| B \|$, for any $A, B \in \mathbb{C}^{nn}$
4. $\| AB \| \leqslant \| A \| \| B \|$, for any $A, B \in \mathbb{C}^{nn}$

There are many matrix norms satisfying the above relations, the most important of which is the matrix Holder norm subordinate to the Holder vector norm, and is given by

$$\| A \|_p = \max_x \frac{\| Ax \|_p}{\| x \|_p}$$

One advantage of choosing the above matrix norm subordinate to the Holder vector norm is that

$$\| Ax \| \leqslant \| A \| \| x \|, \quad \text{for all } A \in \mathbb{C}^{nn} \text{ and } x \in \mathbb{C}^n$$

The proof that $\| A \|_p$ satisfies the above relations is left as an exercise for the reader.

The most popular and widely used norms of the Holder norm are

(a) $$\| A \|_1 = \max_x \frac{\| Ax \|_1}{\| x \|_1} = \max_j \sum_{i=1}^{n} | a_{ij} |$$

Proof

$$\frac{\| Ax \|_1}{\| x \|_1} = \frac{\sum_{i=1}^{n} \left| \sum_{j=1}^{n} a_{ij} x_j \right|}{\sum_{j=1}^{n} | x_j |}$$

$$\leqslant \frac{\sum_{i=1}^{n} \sum_{j=1}^{n} | a_{ij} | | x_j |}{\sum_{j=1}^{n} | x_j |}$$

$$= \frac{\sum_{j=1}^{n} | x_j | \sum_{i=1}^{n} | a_{ij} |}{\sum_{j=1}^{n} | x_j |}$$

Using the Holder inequality explained in Appendix 1, we obtain

$$\frac{\sum_{j=1}^{n} | x_j | \sum_{i=1}^{n} | a_{ij} |}{\sum_{j=1}^{n} | x_j |} \leqslant \frac{\left(\sum_{j=1}^{n} | x_j | \right) \max_j \sum_{i=1}^{n} | a_{ij} |}{\sum_{j=1}^{n} | x_j |} = \max_j \sum_{i=1}^{n} | a_{ij} |$$

Hence we obtain

$$\| A \|_1 \leqslant \max_j \sum_{i=1}^{n} | a_{ij} |$$

Now let k be the number of the column for which $\Sigma_{i=1}^{n} | a_{ij} |$ is maximum.

Choose x such that

$$x_j = 1, \quad \text{for } j = k$$
$$= 0, \quad \text{for } j \neq k$$

Hence

$$\frac{\| Ax \|_1}{\| x \|_1} = \sum_{i=1}^{n} | a_{ik} | = \max_j \sum_{i=1}^{n} | a_{ij} |$$

from which we deduce that

$$\| A \|_1 = \max_j \sum_{i=1}^{n} | a_{ij} |$$

Example

$$A = \begin{bmatrix} 1 & 3 & -1 \\ -2 & 0 & 5 \\ 2 & -4 & 0 \end{bmatrix}. \quad \text{Then } \| A \|_1 = 7.$$

(b) $$\| A \|_2 = \max_x \frac{\| Ax \|_2}{\| x \|_2}$$

This quantity will be calculated later on, when the reader is acquainted with the matrix eigenvalue problem.

(c) $$\| A \|_\infty = \max_x \frac{\| Ax \|_\infty}{\| x \|_\infty} = \max_i \sum_{j=1}^{n} | a_{ij} |$$

Proof

$$\frac{\| Ax \|_\infty}{\| x \|_\infty} = \frac{\max_i \left| \sum_{j=1}^{n} a_{ij} x_j \right|}{\max_j | x_j |}$$

$$\leqslant \frac{\max_i \sum_{j=1}^{n} | a_{ij} | \, | x_j |}{\max_j | x_j |}$$

Using the Holder inequality explained in Appendix 1, we obtain

$$\frac{\max_i \sum_{j=1}^{n} | a_{ij} | \, | x_j |}{\max_j | x_j |} \leqslant \frac{\max_i \left(\sum_{j=1}^{n} | a_{ij} | \cdot \max_j | x_j | \right)}{\max_j | x_j |} = \max_i \sum_{j=1}^{n} | a_{ij} |$$

Hence

$$\| A \|_\infty \leqslant \max_i \sum_{j=1}^{n} | a_{ij} |$$

Now let k be the number of the row for which $\Sigma_{j-1}^{n} \; | a_{ij} |$ is maximum. Choose x such that

$$\begin{aligned} x_j &= 1, \quad \text{for } a_{kj} > 0 \\ &= -1, \quad \text{for } a_{kj} < 0 \end{aligned}$$

Hence

$$\frac{\| Ax \|_\infty}{\| x \|_\infty} = \sum_{j=1}^{n} | a_{kj} | = \max_i \sum_{j=1}^{n} | a_{ij} |$$

from which we conclude that

$$\| A \|_\infty = \max_i \sum_{j=1}^{n} | a_{ij} |$$

Example

$$A = \begin{bmatrix} 0 & -1 & 4 \\ 1 & 2 & -5 \\ -3 & 0 & 4 \end{bmatrix}. \quad \text{Then } \| A \|_\infty = 8.$$

Exercises 2.3

1. Show that $\| x \|_\infty \leqslant \| x \|_2 \leqslant \| x \|_1$
2. Show that $\| x \|_\infty \leqslant \| x \|_1 \leqslant n \| x \|_\infty$
3. Show that $\| x \|_\infty \leqslant \| x \|_2 \leqslant \sqrt{n} \| x \|_\infty$
4. Show that $\dfrac{\| x \|_1}{\sqrt{n}} \leqslant \| x \|_2 \leqslant \| x \|_1$
5. Obtain $\| x \|_1, \| x \|_2, \| x \|_3$ and $\| x \|_\infty$ for $x^T = (1 + i, -2, 5, -3)$
6. Show that $\| I \| = 1$, subordinate to the Holder vector norm.
7. Show that $\| A - B \| \geqslant | \| A \| - \| B \| |$
8. Show that $\| A^n \| \leqslant \| A \|^n$ when n is a positive integer
9. Obtain $\| A \|_1$ and $\| A \|_\infty$ for

$$A = \begin{bmatrix} -1 & 0 & 2 & i \\ 3 & 5 & -1 & 0 \\ 1 & 2 & 0 & -1 \\ 7 & 1 & 2 & -4 \end{bmatrix}, \quad i^2 = -1$$

10. Show that, if U is unitary, $\| U \|_2 = 1$
11. Show that, if U is unitary, $\| UA \|_2 = \| A \|_2$

12. Show that, if U is unitary, $\| UAU^* \|_2 = \| A \|_2$
13. Show that $\| \int x(t)\, dt \| \leqslant \int \| x(t) \|\, dt$
14. Show that $\| \int A(t)\, dt \| \leqslant \int \| A(t) \|\, dt$
15. If a matrix A^{-1} is defined such that $A^{-1}A = I$, show that $\| A^{-1} \| \geqslant \| A \|^{-1}$
16. Show that, if $\| A \| < 1$, then $\| I + A + A^2 + A^3 + \cdots \| \leqslant (1 - \| A \|)^{-1}$
17. If $A = \operatorname{diag}(a_{11}, a_{22}, \ldots, a_{nn})$, show that $\| A \|_2 = \max_i |a_{ii}|$
18. If $\sqrt{\Sigma_{i,j} |a_{ij}|^2}$ is called $\| A \|_E$, where E stands for Euclidean, show that if U is unitary, then $\| UA \|_E = \| A \|_E$. *Hint*: $\Sigma_{i,j} |a_{ij}|^2 = \operatorname{tr}(A^*A)$
19. For a matrix A does the quantity $(\Sigma_{i,j} |a_{ij}|^p)^{1/p}$ represent the norm of A?
20. Show that, for a matrix A, the quantity $n. \max_{i,j} |a_{ij}|$ can be taken as its norm.

2.4. Condition number of a matrix

The *condition number* of a square matrix A, denoted by '$\gamma(A)$', is a non-negative real scalar defined, in connection with the Holder vector norm, as

$$\gamma_p(A) = \max_{u,v} \frac{\| Au \|_p}{\| Av \|_p}, \quad \| u \|_p = \| v \|_p = 1$$

This number has a simple interpretation. Let the surface $\| x \| = 1$ be mapped by the transformation $y = Ax$ onto some surface S. The condition number $\gamma(A)$ is the ratio of the largest to the smallest distances from the origin to points on S. Thus

$$\gamma(A) \geqslant 1$$

To compute $\gamma(A)$ we write it in the following form

$$\gamma(A) = \max_{u,v} \frac{\| Au \|}{\| Av \|} = \left(\max_u \frac{\| Au \|}{\| u \|} \right) \Big/ \left(\min_v \frac{\| Av \|}{\| v \|} \right)$$

$$= \| A \| / \min_v \frac{\| Av \|}{\| v \|}.$$

Now define a matrix A^{-1} such that

$$A^{-1}A = I$$

Hence

$$A^{-1}Av = v$$

And we obtain

$$\| v \| = \| A^{-1}Av \| \leqslant \| A^{-1} \| \cdot \| Av \|$$

Therefore

$$\frac{\| Av \|}{\| v \|} \geqslant \| A^{-1} \|^{-1}$$

From which we obtain

$$\min_{v} \frac{\| Av \|}{\| v \|} = \| A^{-1} \|^{-1}$$

And we finally obtain

$$\gamma(A) = \| A^{-1} \| \, \| A \|$$

If $\gamma(A)$ is large, the matrix A is called *ill-conditioned.* Usually if $|A|$ is small then we find that $\gamma(A)$ is large. A situation like this is not welcomed when solving a set of linear simultaneous equations, since a small change in the coefficients of the equations causes a large displacement in the solution. To see this, consider the following example:

$$0.505x + 0.495y = 1$$
$$x + y = 2$$

which represent two straight lines intersecting at the point

$$x = 1, \quad y = 1$$

Now if the system is altered as follows

$$0.505x + 0.495y = 1$$
$$x + y = 2 + \epsilon,$$

then the new point of intersection becomes

$$x = 1 - 49.5\epsilon, \quad y = 1 + 50.5\epsilon$$

This shows that the system is ill-conditioned (against well-conditioned), in the sense that a small change on the right of magnitude ϵ produces a change in the solution of magnitude 50ϵ approximately. The reason is due to the smallness of the angle between the two lines. For this system of equations

$$A = \begin{bmatrix} 0.505 & 0.495 \\ 1 & 1 \end{bmatrix}, \quad A^{-1} = \begin{bmatrix} 100 & -49.5 \\ -100 & 50.5 \end{bmatrix}$$

and

$$\gamma(A) \approx 300$$

Ill-conditioning may be regarded as an approach towards singularity. It gives inaccurate solutions due to the loss of significant figures during the computation. The condition number comes into many sorts of error analysis as will be explored later on.

Exercises 2.4

1. If $A = \begin{bmatrix} 1 & 2 \\ 3 & 4 \end{bmatrix}$, $A^{-1} = \begin{bmatrix} -2 & 1 \\ 1.5 & -0.5 \end{bmatrix}$, find $\gamma(A)$

2. If $A = \begin{bmatrix} 5 & 1.2 \\ 2 & 0.5 \end{bmatrix}$, $A^{-1} = \begin{bmatrix} 5 & -12 \\ -20 & 50 \end{bmatrix}$, find $\gamma(A)$

3. If $A = \begin{bmatrix} 4 & 1 \\ 3 & 3 \end{bmatrix}$, $A^{-1} = \begin{bmatrix} \frac{1}{3} & \frac{-1}{9} \\ \frac{-1}{3} & \frac{4}{9} \end{bmatrix}$, find $\gamma(A)$

4. Discuss the results in the above three examples, also give an interpretation for $\gamma(A)$ in each.
5. If U is unitary, show that $\gamma(UA) = \gamma(AU) = \gamma(A)$ relative to the norm $\|\cdot\|_2$
6. Show that $\gamma(AB) \leqslant \gamma(A)\gamma(B)$.
7. Show that $\| A^{-1} \| \| A \| \geqslant 1$, if $A^{-1}A = I$, i.e. $\gamma(A) \geqslant 1$ always.
8. If $\dfrac{\| B \|}{\| A \|} = \theta < 1$, show that $\gamma(A + B) \leqslant \dfrac{\gamma(A) + \theta}{1 - \theta}$.
9. Show that $\min\limits_{v} \dfrac{\| Av \|}{\| v \|} = \| A^{-1} \|^{-1}$, where $AA^{-1} = A^{-1}A = I$.
10. If $|A| = 0$, show that $\gamma(A) = \infty$. *Hint*: consider $\min\limits_{v} \dfrac{\| Av \|}{\| v \|}$ taking $v_k = C_{ik}$.

2.5. Inner product of two vectors

The *inner product* of two complex vectors x and y is a scalar quantity denoted by '$\langle x, y \rangle$' and is given by

$$\langle x, y \rangle = x^* y$$

Obviously if both x and y are real then their inner product is real. Instead if each or both of them are complex then their inner product will in general be complex.

Example. Let

$$x = \begin{bmatrix} 1 \\ 2 \\ -1 \end{bmatrix}, \quad y = \begin{bmatrix} 2 \\ -5 \\ 0 \end{bmatrix}. \quad \text{Then } \langle x, y \rangle = 1 \times 2 - 2 \times 5 = -8$$

Example. Let

$$x = \begin{bmatrix} 1 + i \\ -1 \\ 2 - i \end{bmatrix}, \quad y = \begin{bmatrix} 1 \\ i \\ 1 \end{bmatrix}, \quad i^2 = -1. \text{ Then } \langle x, y \rangle = 1 - i - i + 2 + i = 3 - i$$

If the inner product of two vectors is zero, i.e.

$$\langle x, y \rangle = 0$$

the two vectors are said to be *orthogonal*. For example the two vectors

$$x = \begin{bmatrix} 1 \\ 2 \end{bmatrix} \quad \text{and } y = \begin{bmatrix} -2 \\ 1 \end{bmatrix}$$

are orthogonal in the space R^2. Whereas the two vectors

$$x = \begin{bmatrix} 1+i \\ 0 \\ 1-i \end{bmatrix} \quad \text{and } y = \begin{bmatrix} -i \\ i \\ 1 \end{bmatrix}$$

are orthogonal in the space $\mathbb{C}^3$.

The reader will notice that the inner product satisfies the following properties whose proof is left as an exercise for the reader

1. $\langle x, x \rangle = \| x \|_2^2$
2. $\langle x, x \rangle = 0$, if and only if $x = 0$
3. $\langle x, y \rangle = \overline{\langle y, x \rangle}$
4. $\langle x, \alpha y + \beta z \rangle = \alpha \langle x, y \rangle + \beta \langle x, z \rangle$, for any $\alpha, \beta \in \mathbb{C}$, and any $x, y, z \in \mathbb{C}^n$

The relation between the inner product of two vectors and their norms can be obtained using the Holder inequality explained in Appendix 1, as follows:

$$\begin{aligned} |\langle x, y \rangle| &= \left| \sum_{i=1}^{n} \bar{x}_i y_i \right| \\ &\leqslant \sum_{i=1}^{n} |x_i| \, |y_i| \\ &\leqslant \left(\sum_{i=1}^{n} |x_i|^p \right)^{1/p} \left(\sum_{i=1}^{n} |y_i|^q \right)^{1/q} \\ &= \| x \|_p \| y \|_q \end{aligned}$$

Hence

$$|\langle x, y \rangle| = \| x \|_p \| y \|_q \cos \theta$$

where θ takes a physical meaning if $p = q = 2$, and x and $y \in \mathbb{R}^3$. Hence the natural definition:

$$\cos \theta = \frac{\langle x, y \rangle}{\| x \|_2 \| y \|_2} (-\Pi < \theta < \Pi)$$

The definition of orthogonality between two vectors can be well extended to matrices. Hence two matrices A and B are orthogonal on each other if the inner product of any two vectors or any two columns belonging to both matrices is zero, i.e.

$$AB^* = 0$$

Examples of such matrices are the *cut-set* and *tie-set* of any electrical circuit. For example consider the circuit below and write the two Kirchhoff's laws in the following matrix forms.

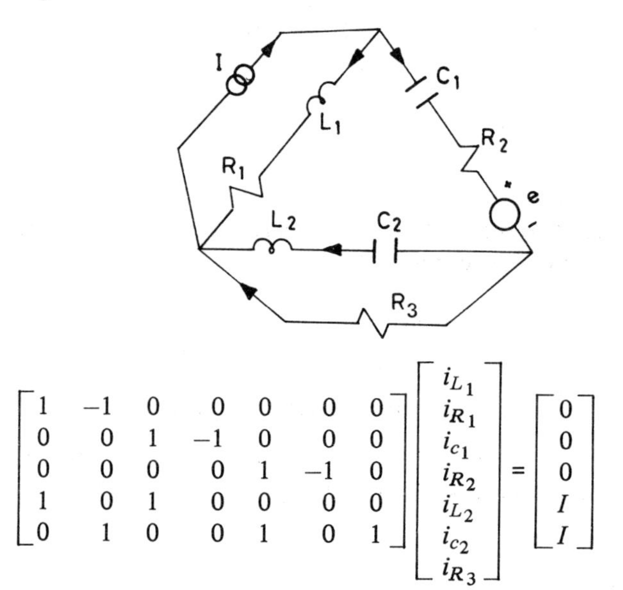

$$\begin{bmatrix} 1 & -1 & 0 & 0 & 0 & 0 & 0 \\ 0 & 0 & 1 & -1 & 0 & 0 & 0 \\ 0 & 0 & 0 & 0 & 1 & -1 & 0 \\ 1 & 0 & 1 & 0 & 0 & 0 & 0 \\ 0 & 1 & 0 & 0 & 1 & 0 & 1 \end{bmatrix} \begin{bmatrix} i_{L_1} \\ i_{R_1} \\ i_{c_1} \\ i_{R_2} \\ i_{L_2} \\ i_{c_2} \\ i_{R_3} \end{bmatrix} = \begin{bmatrix} 0 \\ 0 \\ 0 \\ I \\ I \end{bmatrix}$$

and

$$\begin{bmatrix} 1 & 1 & -1 & -1 & -1 & -1 & 0 \\ 0 & 0 & 0 & 0 & 1 & 1 & -1 \end{bmatrix} \begin{bmatrix} V_{L_1} \\ V_{R_1} \\ V_{c_1} \\ V_{R_2} \\ V_{L_2} \\ V_{c_2} \\ V_{R_3} \end{bmatrix} = \begin{bmatrix} e \\ 0 \end{bmatrix}$$

It can be noticed that the coefficient matrices are orthogonal on each other.

2.6. Quadratic forms

A *quadratic form* is an inner product of a vector x over matrix Q. It takes the form

$$\langle x, Qx \rangle = x^*Qx.$$

So if x and Q are both real then the quadratic form is real and if they are complex, then it becomes generally complex.

Example

$$x = \begin{bmatrix} 1 \\ i \\ 0 \end{bmatrix}, \quad Q = \begin{bmatrix} 1 & -1 & 0 \\ 0 & 1 & i \\ 2 & -2 & -1 \end{bmatrix}, \quad i^2 = -1. \text{ Then } x^*Qx = 2 - i$$

The quadratic form appears frequently in physics in the form of energy stored or power dissipated in a system.

Example. Find the energy stored in the circuit shown below:

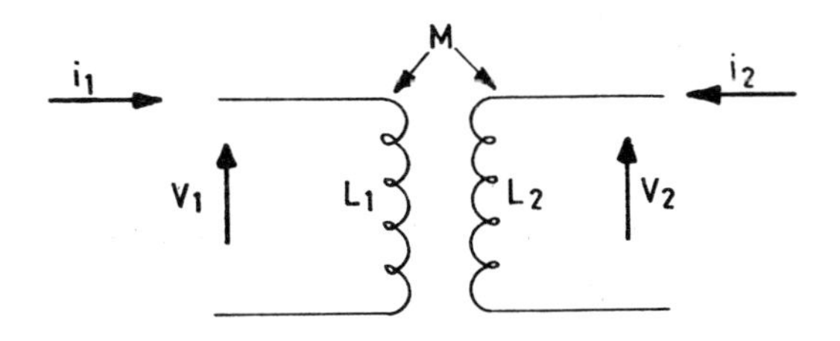

$$\text{Energy stored} = \int \langle v, i \rangle \, dt$$

$$= \int \left\langle \begin{bmatrix} L_1 & M \\ M & L_2 \end{bmatrix} \begin{bmatrix} \dot{i}_1 \\ \dot{i}_2 \end{bmatrix}, \begin{bmatrix} i_1 \\ i_2 \end{bmatrix} \right\rangle dt$$

$$= \tfrac{1}{2} L_1 i_1^2 + \tfrac{1}{2} L_2 i_2^2 + M i_1 i_2$$

Example. Find the kinetic energy for the rigid body rotating, about three axes in it, with angular velocity w_x, w_y, w_z

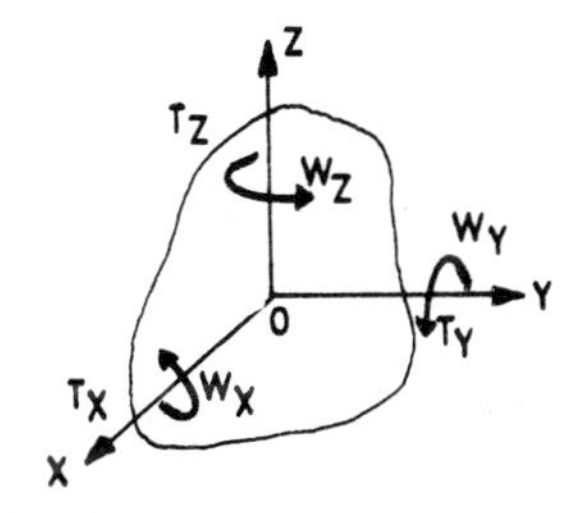

$$\text{Kinetic energy} = \int \langle T, w \rangle \, dt$$

$$= \int \left\langle \begin{bmatrix} I_{xx} & -I_{xy} & -I_{xz} \\ -I_{xy} & I_{yy} & -I_{yz} \\ -I_{xz} & -I_{yz} & I_{zz} \end{bmatrix} \begin{bmatrix} \dot{w}_x \\ \dot{w}_y \\ \dot{w}_z \end{bmatrix}, \begin{bmatrix} w_x \\ w_y \\ w_z \end{bmatrix} \right\rangle dt$$

$$= \tfrac{1}{2} I_{xx} w_x^2 + \tfrac{1}{2} I_{yy} w_y^2 + \tfrac{1}{2} I_{zz} w_z^2 - I_{xy} w_y w_x - I_{xz} w_x w_z - I_{yz} w_y w_z$$

Example. Represent the surface

$$2x^2 + 3y^2 + z^2 + 2xy + xz = 1$$

in a quadratic form

$$2x^2 + 3y^2 + z^2 + 2xy + xz = [x \quad y \quad z] \begin{bmatrix} 2 & 1 & 0.5 \\ 1 & 3 & 0 \\ 0.5 & 0 & 1 \end{bmatrix} \begin{bmatrix} x \\ y \\ z \end{bmatrix} = 1$$

Example. Find the power dissipated in the n-port network

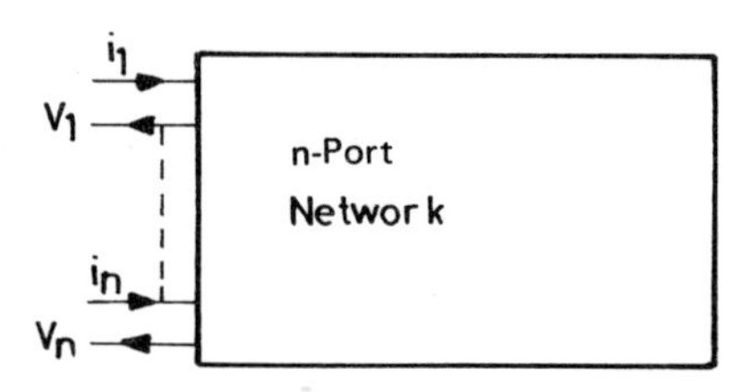

The power dissipated in the ith port is equal to the real part of $\bar{v}_i i_i$ where v_i and i_i are r.m.s. values. Hence the overall dissipated power is equal to Re $\langle v, i \rangle$ where

$$v = \begin{bmatrix} v_1 \\ v_2 \\ \vdots \\ v_n \end{bmatrix} \quad \text{and } i = \begin{bmatrix} i_1 \\ i_2 \\ \vdots \\ i_n \end{bmatrix}$$

But

$$v = Zi$$

where Z is the impedance matrix of the n-port. So we obtain overall dissipated power $= \text{Re}\langle Zi, i \rangle = \frac{1}{2} i^*(Z + Z^*)\, i$.

Exercises 2.6

1. If $x^T = [1 \quad i \quad -3]$, $y = [-i, 1+i, 7]$ and $Q = \begin{bmatrix} 1 & i & -1 \\ -1 & 0 & 3 \\ 4 & 0 & 1+i \end{bmatrix}$, $i^2 = -1$
obtain $\langle x, y \rangle$, $\langle y, x \rangle$, $\langle x, Qy \rangle$ and $\langle Qx, y \rangle$.
2. If Q is Hermitian show that $\langle x, Qx \rangle = \langle Qx, x \rangle$.
3. Show that $x^T Q x = x^T Q^T x$.
4. Show that if A is skew-symmetric, then $x^T A x = 0$. Demonstrate your result by an example.
5. Does $x^T A x = x^T B x$ for all x imply that $A = B$?
6. Under what condition does $x^T A x = 0$ for all x?
7. Show that $|\langle x, Qx \rangle| \leqslant \| x \|_2^2 \| Q \|_2$.
8. Show that $|\langle x, Qx \rangle| \leqslant \| x \|_p \| x \|_q \| Q \|_q$.
9. If Q is Hermitian, show that $\langle x, Qx \rangle$ is real.
10. If Q is skew-Hermitian, show that $\langle x, Qx \rangle$ is pure imaginary.
11. Show that $\text{Re}\langle x, Qx \rangle = \frac{1}{2}\langle x, (Q + Q^*)x \rangle$.
12. Show that imaginary $\langle x, Qx \rangle = \frac{1}{2}\langle x, (Q - Q^*)x \rangle$.
13. Does $\langle x, x \rangle = \langle y, y \rangle$ imply that $x = y$?
14. Show that $\langle x + y, x + y \rangle = \| x \|_2^2 + \| y \|_2^2 + \langle x, y \rangle + \langle y, x \rangle$.
15. Show that $|\langle x + y, x + y \rangle| \leqslant (\| x \|_2 + \| y \|_2)^2$.
16. Show that $|\langle x, y \rangle + \langle y, x \rangle| \leqslant 2 \| x \|_2 \| y \|_2$.
17. Show that $x^T A_1 x = x^T A_2 x$, where A_1 and A_2 are symmetric matrices; implies that $A_1 = A_2$ (this proves that the symmetric matrix of a quadratic form is unique).

CHAPTER THREE

THEORY OF LINEAR EQUATIONS

3.1. Linear dependence and independence over a field

We begin our study of the linear dependence and independence of a set of vectors $x^1, x^2, \ldots, x^m$ by the following definition.

DEFINITION 1. If there exists a set of m vectors $x^1, x^2, \ldots, x^m$, and a set of m scalars $c_1, c_2, \ldots, c_m$ with a linear relation of the form

$$c_1 x^1 + c_2 x^2 + \cdots + c_m x^m = 0$$

then the vectors $x^1, x^2, \ldots, x^m$ are called *linearly independent* if the above relation implies that

$$c_1 = c_2 = \cdots = c_m = 0$$

But if one or more of the coefficients $c_1, c_2, \ldots, c_m$ can be non-zero, then the vectors are called *linearly dependent.*

If the coefficients are real the vectors are said to be dependent over the field of real numbers; likewise if the coefficients are complex the vectors are said to be dependent over the field of complex numbers.

Example

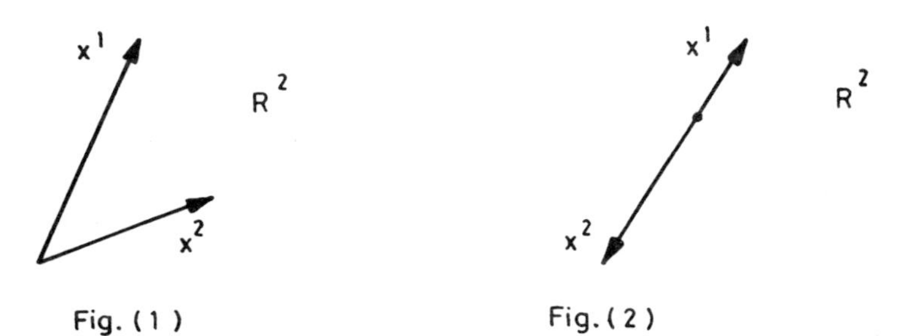

Fig. (1) Fig. (2)

In figure 1 the vectors are linearly independent in the sense that there is no linear relation of the type

$$c_1 x^1 + c_2 x^2 = 0$$

In other words if such a relation did exist, then $c_1 = c_2 = 0$. On the other hand the vectors of figure 2 are linearly dependent for there exists a relation of the type

$$c_1x^1 + c_2x^2 = 0, \quad c_1, c_2 \neq 0$$

Example. Test the dependence of the following vectors

$$x^1 = \begin{bmatrix} 1 \\ 1 \end{bmatrix}, \quad x^2 = \begin{bmatrix} 0 \\ 2 \end{bmatrix}, \quad x^1 \text{ and } x^2 \in \mathbb{R}^2$$

We form the linear relation

$$c_1 \begin{bmatrix} 1 \\ 1 \end{bmatrix} + c_2 \begin{bmatrix} 0 \\ 2 \end{bmatrix} = \begin{bmatrix} 0 \\ 0 \end{bmatrix}$$

From which we obtain $c_1 = c_2 = 0$; hence x^1 and x^2 are linearly independent.

Example. Test the dependence of the following vectors

$$x^1 = \begin{bmatrix} 1 \\ i \end{bmatrix}, \quad x^2 = \begin{bmatrix} i \\ -1 \end{bmatrix}, \quad i^2 = -1, \quad x^1 \text{ and } x^2 \in \mathbb{C}^2$$

We form the linear relation

$$c_1 \begin{bmatrix} 1 \\ i \end{bmatrix} + c_2 \begin{bmatrix} i \\ -1 \end{bmatrix} = \begin{bmatrix} 0 \\ 0 \end{bmatrix}$$

There are no non-zero real scalars c_1 and c_2 to make the vectors dependent; in other words

$$c_1 + ic_2 = 0$$
$$c_1 i - c_2 = 0$$

implies that

$$c_1 = c_2 = 0$$

Therefore the vectors are called independent over the field of real numbers. However there exist non-zero complex quantities c_1 and c_2 such that

$$c_1x^1 + c_2x^2 = 0$$

in other words such that x^1 and x^2 are linearly dependent. For example one can choose

$$c_1 = 1, \quad c_2 = i$$

The vectors x^1 and x^2 are hence said to be dependent over the field of complex numbers.

Now we are in a position to list some theorems on the linear dependence and independence of a set of vectors $x^1, x^2, \ldots, x^m$. All these theorems rely on the definition of linear dependence and independence cited before. A résumé of these theorems can be found in Kreko (1962), p. 117.

1. Any set of vectors containing the zero vector is linearly dependent.

Proof. If in the linear relation

$$c_1x^1 + c_2x^2 + \cdots + c_mx^m = 0$$

we have

$$c_1 = c_2 = \cdots = c_{m-1} = 0$$

but $c_m \neq 0$, we still obtain

$$0x^1 + 0x^2 + \cdots + 0x^{m-1} + c_mx^m = 0,$$

if $x^m = 0$. Hence $x^1, \ldots, x^m$ are linearly dependent.

2. Any non-empty subset of a set of linearly independent vectors is itself a linearly independent system.

Proof. Assume that the vectors $x^1, x^2, \ldots, x^m$ are linearly independent. If we cancel one of them, say x^1, the remaining system $x^2, \ldots, x^m$ is linearly independent. For if the equation

$$c_2x^2 + c_3x^3 + \cdots + c_mx^m = 0$$

has a non-trivial solution for $c_2, \ldots, c_m$, then so does

$$c_1x^1 + c_2x^2 + \cdots + c_mx^m = 0$$

if $c_1 = 0$. But this contradicts the assumption that $x^1, x^2, \ldots, x^m$ are linearly independent. Therefore, if we remove a vector from a system of linearly independent vectors, the remaining vectors are still linearly independent. If we remove another vector from the remaining ones, we still have a linearly independent system and so on, which completes the proof.

3. If the vectors $x^1, x^2, \ldots, x^m$ are linearly dependent, at least one of them can be written as a linear combination of the others.

Proof. As $x^1, x^2, \ldots, x^m$ are linearly dependent, there exists a set of coefficients $c_1, \ldots, c_m$, at least one of them non-zero, such that

$$c_1x^1 + c_2x^2 + \cdots + c_mx^m = 0$$

Let c_i be such a non-zero coefficient. Hence we obtain

$$x_i = -\frac{1}{c_i}(c_1x^1 + \cdots + c_mx^m)$$

4. If the vector y can be written as a linear combination of $x^1, \ldots, x^m$ the set of vectors $y, x^1, \ldots, x^m$ form a linearly dependent set.

Proof

$$y = c_1x^1 + \cdots + c_mx^m$$

Hence

$$-y + c_1x^1 + \cdots + c_mx^m = 0$$

The set $y, x^1, \ldots, x^m$ is linearly dependent, for at least the coefficient of y is non-zero.

Now we can study further a vector system, which is a finite set of vectors in the vector space V, and make the following definition.

DEFINITION 2. The rank of the set of vectors $x^1, x^2, \ldots, x^m$ is equal to the maximum number of linearly independent vectors in the set. This is equal to the total number of vectors in the set minus the number of linear relations existing among them.

The next theorems discuss properties of the rank of a vector system.

5. If r is the rank of a vector system, every vector in the system can be written as a linear combination of any r linearly independent vectors of the system, and this representation is unique.

Proof. Assume that r out of m vectors are linearly independent; call them $x^1, x^2, \ldots, x^r$. Any vector x^k from the remaining $(m - r)$ vectors is linearly dependent on $x^1, \ldots, x^r$ by definition of the rank; and we obtain

$$c_kx^k + c_1x^1 + \cdots + c_rx^r = 0$$

i.e.

$$x^k = -\frac{1}{c_k}(c_1x^1 + \cdots + c_rx^r), \quad c_k \neq 0$$
$$= \alpha_1x^1 + \cdots + \alpha_rx^r$$

Now for the second part of the theorem, assume that x^k takes a different representation in $x^1, \ldots, x^r$ such as the following

$$x^k = \tilde{\alpha}_1x^1 + \cdots + \tilde{\alpha}_rx^r$$

Subtracting the last two equations we obtain

$$0 = (\alpha_1 - \tilde{\alpha}_1)x^1 + \cdots + (\alpha_r - \tilde{\alpha}_r)x^r$$

But $x^1, \ldots, x^r$ are linearly independent; hence we obtain

$$\alpha_1 = \tilde{\alpha}_1, \ldots, \alpha_r = \tilde{\alpha}_r,$$

and the proof is complete.

6. If a vector that can be written as a linear combination of the other vectors in the system is removed from the given vector system, the rank of the system remains unchanged.

Proof. Let the vector system be $x^1, \ldots, x^r, x^{r+1}, \ldots, x^m$ of rank r, and whose first r vectors are linearly independent. Let us express x^r as a linear combination of the rest, i.e.

$$x^r = c_1x^1 + \cdots + c_{r-1}x^{r-1} + c_{r+1}x^{r+1} + \cdots + c_mx^m$$

The validity of the above equation stems from the fact that all vectors are linearly dependent on $x^1, \ldots, x^r$. Now assume that the rank of the system without x^r is $r - 1$. According to the last theorem all the x^i (for $i = r + 1, \ldots, m$) can be written as a linear combination of the x^j (for $j = 1, \ldots, r - 1$), i.e.

$$x^i = \alpha_{i,1}x^1 + \cdots + \alpha_{i,r-1}x^{r-1}, \quad i = r + 1, \ldots, m$$

And consequently

$$x^r = (c_1 + c_{r+1}\alpha_{r+1,1} + \cdots + c_m\alpha_{m,1})\,x^1 + \cdots + \\ + (c_{r-1} + c_{r+1}\alpha_{r+1,r-1} + \cdots + c_m\alpha_{m,r-1})\,x^{r-1}$$

This shows that x^r is linearly dependent on $x^1, \ldots, x^{r-1}$, which contradicts the assumption that the first r vectors are linearly independent. Hence, even in this case the rank is preserved and the proof is complete.

7. If a vector system is changed by adding a vector which can be represented as a linear combination of vectors already in the system, the rank of the system remains unchanged.

Proof. Given a vector system $x^1, x^2, \ldots, x^m$, let y be a vector which can be represented as a linear combination of these vectors. Consider the vector system $x^1, x^2, \ldots, x^m, y$. According to the last theorem, removing y will not alter the rank, hence both systems have the same rank and the theorem is proved.

8. If the vectors of the system $x^1, x^2, \ldots, x^k$ can be represented as a linear combination of the vectors $y^1, y^2, \ldots, y^p$ in the system, the rank of the system $x^1, x^2, \ldots, x^k$ is at most equal to p.

Proof. Consider the system

$$x^1, x^2, \ldots, x^k, y^1, y^2, \ldots, y^p.$$

According to the last theorem, the rank of the above system is equal to that of the system $y^1, \ldots, y^p$. As the rank of this set is at most equal to p, the result follows immediately.

3.2. Dimension and basis

If in a vector space V there is a maximum number of linearly independent vectors, this number is called the *dimension* of the space. If m is such a number, then any vector system consisting of m linearly independent vectors in the space is called a *basis.* The vectors in a basis are called base vectors.

A fundamental property of a basis is that any vector in the vector space can be represented as a linear combination of the base vectors, and this representation is unique.

To prove this proposition, let $x^1, \ldots, x^m$ be a basis of the space of dimension m. Let y be any other vector in the same vector space. The vector system $x^1, x^2, \ldots, x^m$, y has a rank equal to m, since there are no more than m linearly independent vectors in the vector space, also since $x^1, x^2, \ldots, x^m$ is of rank m, then according to Theorem 5 in Section 3.1, y can be represented uniquely in terms of the basis $x^1, x^2, \ldots, x^m$, i.e.

$$y = \sum_{i=1}^{m} c_i x^i$$

Example. Let

$$x^1 = \begin{bmatrix} 1 \\ 2 \end{bmatrix}, \quad x^2 = \begin{bmatrix} 2 \\ 3 \end{bmatrix}$$

be a basis in $\mathbb{R}^2$. Any vector y, for example

$$y = \begin{bmatrix} 5 \\ 8 \end{bmatrix},$$

can be written as

$$\begin{bmatrix} 5 \\ 8 \end{bmatrix} = c_1 \begin{bmatrix} 1 \\ 2 \end{bmatrix} + c_2 \begin{bmatrix} 2 \\ 3 \end{bmatrix}$$

giving $c_1 = 1, c_2 = 2$.

3.3. Orthogonality and biorthogonality of vectors

A set of vectors $x^1, x^2, \ldots, x^m$ are called *orthogonal* to each other if

$$\langle x^j, x^i \rangle = 0, \quad \text{for all } i \neq j.$$

Example. The set of vectors

$$\begin{bmatrix} 1 \\ 0 \\ 0 \end{bmatrix}, \quad \begin{bmatrix} 0 \\ 1 \\ -2 \end{bmatrix}, \quad \begin{bmatrix} 0 \\ 2 \\ 1 \end{bmatrix}$$

are orthogonal in $\mathbb{R}^3$.
The definition of orthogonality enables us to deduce the following theorem.

THEOREM 1. Non-zero orthogonal vectors are linearly independent.

Proof. Let $x^1, x^2, \ldots, x^m$ be such vectors. Form the linear relation

$$c_1 x^1 + c_2 x^2 + \cdots + c_m x^m = 0,$$

and proceed to prove that

$$c_1 = c_2 = \cdots = c_m = 0.$$

Taking the inner product with x^1, we obtain

$$c_1\langle x^1, x^1 \rangle = 0$$

But

$$\langle x^1, x^1 \rangle = \| x^1 \|_2^2 > 0$$

as $x^1 \neq 0$; hence

$$c_1 = 0$$

Similarly we can show that

$$c_2 = c_3 = \cdots = c_m = 0$$

and the proof is complete.

The concept of orthogonality of vectors facilitates the expansion of any vector y in terms of orthogonal vectors. For let $x^1, x^2, \ldots, x^n$ be a set of non-zero orthogonal vectors in a space of dimension n, and y be any vector in the same space that we desire to expand in the form

$$y = \sum_{i=1}^{n} c_i x^i$$

Taking the inner product with x^j, we obtain

$$c_j = \frac{\langle x^j, y \rangle}{\langle x^j, x^j \rangle}$$

If in addition we have

$$\langle x^j, x^j \rangle = 1,$$

the vector system $x^1, \ldots, x^n$ is called an *orthonormal system* and the vectors $x^1, \ldots, x^n$ are called an *orthonormal basis.*

Example. Let

$$x^1 = \begin{bmatrix} 1 \\ 2 \\ 0 \end{bmatrix}, \quad x^2 = \begin{bmatrix} 2 \\ -1 \\ -1 \end{bmatrix}, \quad x^3 = \begin{bmatrix} -2 \\ 1 \\ -5 \end{bmatrix}, \quad y = \begin{bmatrix} -3 \\ 0 \\ 4 \end{bmatrix}$$

Obtain c_1, c_2 and c_3 such that $y = c_1x^1 + c_2x^2 + c_3x^3$

Using the above technique we get:

$$c_1 = \frac{\langle y, x^1 \rangle}{\langle x^1, x^1 \rangle} = \frac{-3}{5}$$

$$c_2 = \frac{\langle y, x^2 \rangle}{\langle x^2, x^2 \rangle} = \frac{-10}{6}$$

$$c_3 = \frac{\langle y, x^3 \rangle}{\langle x^3, x^3 \rangle} = \frac{-14}{30}$$

Example. Transform the set of orthogonal vectors

$$x^1 = \begin{bmatrix} 1 \\ 2 \\ 0 \end{bmatrix}, \quad x^2 = \begin{bmatrix} 2 \\ -1 \\ -1 \end{bmatrix}, \quad x^3 = \begin{bmatrix} -2 \\ 1 \\ -5 \end{bmatrix}$$

into an orthonormal basis.

Dividing by the norm of each we obtain the orthonormal basis as follows:

$$x_n^1 = \begin{bmatrix} \frac{1}{\sqrt{5}} \\ \frac{2}{\sqrt{5}} \\ 0 \end{bmatrix}, \quad x_n^2 = \begin{bmatrix} \frac{2}{\sqrt{6}} \\ \frac{-1}{\sqrt{6}} \\ \frac{-1}{\sqrt{6}} \end{bmatrix}, \quad x_n^3 = \begin{bmatrix} \frac{-2}{\sqrt{30}} \\ \frac{1}{\sqrt{30}} \\ \frac{-5}{\sqrt{30}} \end{bmatrix}$$

A very simple example of an orthonormal basis in the space $\mathbb{R}^n$ is the n-axis $e^1, \ldots, e^n$ where

$$e^1 = \begin{bmatrix} 1 \\ 0 \\ \vdots \\ 0 \end{bmatrix}, \quad e^2 = \begin{bmatrix} 0 \\ 1 \\ \vdots \\ 0 \end{bmatrix}, \ldots, e^n = \begin{bmatrix} 0 \\ 0 \\ \vdots \\ 1 \end{bmatrix}$$

If $x^1, \ldots, x^n$ are not orthogonal, but only independent, then to solve for the coefficients $c_1, c_2, \ldots, c_n$, the equations

$$y = \sum_{i=1}^{n} c_i x^i$$

is to solve n linear simultaneous equations in n unknowns. Until this is studied in later sections of this chapter, we use the concept of biorthogonality, which enables us to obtain $c_1, c_2, \ldots, c_n$ if the biorthogonal basis of $x^1, x^2, \ldots, x^n$ is known. We start by making the following definition.

DEFINITION. Two sets of vectors $x^1, x^2, \ldots, x^n$ and $y^1, y^2, \ldots, y^n$ are called *biorthogonal* if

$$\begin{aligned} \langle x^j, y^i \rangle &= 0, \quad \text{for } i \neq j \\ &\neq 0, \quad \text{for } i = j \end{aligned}$$

In the literature, we find some authors using the term *reciprocal* instead of *biorthogonal*. The concept of biorthogonality enables us to establish the following theorem.

THEOREM 2. If the two sets of vectors $x^1, x^2, \ldots, x^n$ and $y^1, y^2, \ldots, y^n$ are biorthogonal, the vectors of each set are linearly independent.

Proof. We prove here that $x^1, x^2, \ldots, x^n$ are linearly independent. To prove that $y^1, y^2, \ldots, y^n$ are linearly independent is similar. Form the relation

$$c_1 x^1 + c_2 x^2 + \cdots + c_n x^n = 0$$

and proceed to prove that

$$c_1 = c_2 = \cdots = c_n = 0$$

Taking the inner product with y^1, we obtain

$$c_1 \langle y^1, x^1 \rangle + c_2 \langle y^1, x^2 \rangle + \cdots + c_n \langle y^1, x^n \rangle = 0$$

All the terms from the second term up to the last are zero according to the biorthogonality condition. Hence

$$c_1 \langle y^1, x^1 \rangle = 0$$

However

$$\langle y^1, x^1 \rangle \neq 0$$

according to the definition of biorthogonality. Hence

$$c_1 = 0$$

Similarly one can show that

$$c_2 = \cdots = c_n = 0$$

and the proof is complete.

Example

$$x^1 = \begin{bmatrix} 1 \\ 2 \end{bmatrix}, \quad x^2 = \begin{bmatrix} 3 \\ -4 \end{bmatrix}, \quad y^1 = \begin{bmatrix} -4 \\ -3 \end{bmatrix}, \quad y^2 = \begin{bmatrix} -2 \\ 1 \end{bmatrix}$$

The reader can check that the two sets x^1, x^2 and y^1, y^2 are biorthogonal.

The concept of biorthogonality facilitates the expansion of any vector v in terms of any set of linearly independent vectors $x^1, \ldots, x^n$ lying in a space of dimension n if their reciprocal vectors $y^1, \ldots, y^n$ are known. For let

$$v = \sum_{i=1}^{n} c_i x^i$$

To obtain c_j we take the inner product of the above equation with y^j giving

$$c_j = \frac{\langle y^j, v\rangle}{\langle y^j, x^j\rangle}$$

Example. Let

$$x^1 = \begin{bmatrix} 1 \\ 2 \\ 0 \end{bmatrix}, \quad x^2 = \begin{bmatrix} 1 \\ 1 \\ 2 \end{bmatrix}, \quad x^3 = \begin{bmatrix} 0 \\ 1 \\ 2 \end{bmatrix}, \quad v = \begin{bmatrix} 1 \\ -1 \\ 1 \end{bmatrix}$$

$$y^1 = \begin{bmatrix} 0 \\ 2 \\ -1 \end{bmatrix}, \quad y^2 = \begin{bmatrix} 4 \\ -2 \\ 1 \end{bmatrix}, \quad y^3 = \begin{bmatrix} -4 \\ 2 \\ 1 \end{bmatrix}$$

Then if

$$v = c_1x^1 + c_2x^2 + c_3x^3$$

we have

$$c_1 = \frac{\langle y^1, v\rangle}{\langle y^1, x^1\rangle} = \frac{-3}{4}$$

$$c_2 = \frac{\langle y^2, v\rangle}{\langle y^2, x^2\rangle} = \frac{7}{4}$$

$$c_3 = \frac{\langle y^3, v\rangle}{\langle y^3, x^3\rangle} = \frac{-5}{4}$$

3.4. The Grammian

Until now we have not given a method for testing the linear independence of vectors, except by relying upon the definition of independence itself. We have set out the linear relation

$$c_1x^1 + c_2x^2 + \cdots + c_mx^m = 0$$

and then proceeded to show that

$$c_1 = c_2 = \cdots = c_m = 0$$

Although the method seems simple, to show that the coefficients are all zero is a very tedious exercise. The method to be explained in this section gives a straightforward answer to whether the vectors are linearly independent or not.

We define the *Grammian matrix* or *Grammian* for a set of vectors $x^1, x^2, \ldots, x^m$ as follows:

$$\underset{m,m}{G} = \begin{bmatrix} \langle x^1,x^1 \rangle & \langle x^1,x^2 \rangle & \dots & \langle x^1,x^m \rangle \\ \langle x^2,x^1 \rangle & \langle x^2,x^2 \rangle & \dots & \\ \vdots & & & \\ \langle x^m,x^1 \rangle & & \dots & \langle x^m,x^m \rangle \end{bmatrix}$$

Therefore if A is the matrix whose columns are $x^1, x^2, \dots, x^m$, it appears directly that

$$G = A^*A$$

from which we conclude that the Grammian G is Hermitian and consequently its determinant is real. Moreover, the determinant of G can be shown to be greater than or equal to zero, and the reader should do this as an exercise.

Now we proceed to give the test of linear independence for a set of vectors $x^1, x^2, \dots, x^m$ as a result of the following theorem.

THEOREM 1. If the determinant of the Grammian of a set of vectors $x^1, x^2, \dots, x^m$ is greater than zero, the vectors are linearly independent.

Proof. Form the linear relation

$$c_1x^1 + c_2x^2 + \cdots + c_mx^m = 0,$$

and proceed to prove that

$$c_1 = c_2 = \cdots = c_m = 0.$$

Taking the inner product of the above equation first with x^1, then x^2, until x^m, we obtain

$$\begin{bmatrix} \langle x^1,x^1 \rangle & \langle x^1,x^2 \rangle & \dots & \langle x^1,x^m \rangle \\ \langle x^2,x^1 \rangle & \langle x^2,x^2 \rangle & \dots & \langle x^2,x^m \rangle \\ \vdots & & & \\ \langle x^m,x^1 \rangle & & \dots & \langle x^m,x^m \rangle \end{bmatrix} \begin{bmatrix} c_1 \\ c_2 \\ \vdots \\ c_m \end{bmatrix} = 0$$

Hence if $|G| > 0$, then according to Exercise 2.1.14, we obtain

$$c_1 = c_2 = \cdots = c_m = 0$$

and the proof is complete.

THEOREM 2. A necessary and sufficient condition that $x^1, x^2, \dots, x^m$ are linearly dependent is that their Grammian determinant is zero.

Proof. Necessity can be established from Exercise 2.1.14. To prove sufficiency we notice from Exercise 2.1.14 that if the Grammian determinant is zero we have two possibilities: either the coefficients are zero which is trivial, or the coefficients may not be zero, i.e. there exists a non-trivial solution for the coefficients, which completes the proof.

A special case arises when $x^1, \ldots, x^m$ lie in the space $\mathbb{R}^m$ or $\mathbb{C}^m$. In this case the matrix A, whose columns are the vectors $x^1, \ldots, x^m$, will be square, and we get the following results which are special cases of Theorems 1 and 2 above.

1. If $|A| \neq 0$, then $x^1, \ldots, x^m$ are linearly independent
2. A necessary and sufficient condition that $x^1, \ldots, x^m$ are linearly dependent is that $|A| = 0$

The proof of the above results is direct, and so it is left as an exercise for the reader.

One important result of the Grammian is that if $x^1, \ldots, x^m$ lie in the space $\mathbb{R}^n$ or $\mathbb{C}^n$ with $m > n$, then $x^1, \ldots, x^m$ are linearly dependent. In other words, one cannot find in the Euclidean space of dimension n more than n linearly independent vectors. See Exercise 2.1.18.

3.5. The Gram–Schmidt process

Sometimes it becomes useful to obtain a set of orthogonal vectors from a set of independent vectors. One such use is met when dealing with the eigenvalue problem of Hermitian matrices as will be seen in the next chapter. The method of generating n orthogonal vectors from n linearly independent vectors is called the Gram–Schmidt process and is as follows.

Let $x^1, x^2, \ldots, x^n$ be a set of linearly independent vectors and it is required to obtain from them a set of orthogonal vectors $y^1, y^2, \ldots, y^n$. We choose the last vectors as follows:

$$y^1 = x^1$$
$$y^2 = x^2 + \alpha x^1$$

and proceed to find α such that y^2 and y^1 are mutually orthogonal. Taking the inner product with y^1, we obtain

$$\langle y^1, y^2 \rangle = \langle x^1, y^2 \rangle = 0 = \langle x^1, x^2 \rangle + \alpha \langle x^1, x^1 \rangle$$

Hence

$$\alpha = -\frac{\langle x^1, x^2 \rangle}{\langle x^1, x^1 \rangle}$$

Choose y^3 in the following manner:

$$y^3 = x^3 + c_2 x^2 + c_1 x^1$$

and proceed to find c_1 and c_2 such that y^3, y^2 and y^1 are mutually orthogonal. Taking the inner product with y^1, we obtain

$$\langle y^1, y^3 \rangle = \langle x^1, y^3 \rangle = 0 = \langle x^1, x^3 \rangle + c_2 \langle x^1, x^2 \rangle + c_1 \langle x^1, x^1 \rangle$$

Now if y^3 is made orthogonal on x^2, then y^3 will consequently be orthogonal on

y^2. Hence taking the inner product with x^2 gives

$$\langle x^2, y^3 \rangle = 0 = \langle x^2, x^3 \rangle + c_2 \langle x^2, x^2 \rangle + c_1 \langle x^2, x^1 \rangle$$

The above two equations can be put in the following form:

$$\begin{bmatrix} \langle x^1, x^1 \rangle & \langle x^1, x^2 \rangle \\ \langle x^2, x^1 \rangle & \langle x^2, x^2 \rangle \end{bmatrix} \begin{bmatrix} c_1 \\ c_2 \end{bmatrix} = - \begin{bmatrix} \langle x^1, x^3 \rangle \\ \langle x^2, x^3 \rangle \end{bmatrix}$$

The matrix on the left-hand side has a non-zero determinant, for it is the Grammian of the two linearly independent vectors x^1 and x^2. Therefore the coefficients c_1 and c_2 can be calculated uniquely.

The process can be prolonged similarly on y^4 if it is defined as

$$y^4 = x^4 + c_1 x^1 + c_2 x^2 + c_3 x^3$$

and so on until we take y^n to be

$$y^n = x^n + c_1 x^1 + c_2 x^2 + \cdots + c_{n-1} x^{n-1}$$

The coefficients $c_1, c_2, \ldots, c_{n-1}$ are obtained by solving the linear equations

$$\begin{bmatrix} \langle x^1, x^1 \rangle & \ldots & \langle x^1, x^{n-1} \rangle \\ \langle x^2, x^1 \rangle & \ldots & \\ \vdots & & \\ \langle x^{n-1}, x^1 \rangle & \ldots & \langle x^{n-1}, x^{n-1} \rangle \end{bmatrix} \begin{bmatrix} c_1 \\ c_2 \\ \vdots \\ c_{n-1} \end{bmatrix} = - \begin{bmatrix} \langle x^1, x^n \rangle \\ \\ \vdots \\ \langle x^{n-1}, x^n \rangle \end{bmatrix}$$

The matrix of the left-hand side is nonsingular for it is the Grammian of the set of linearly independent vectors $x^1, \ldots, x^{n-1}$.

Exercises 3.5

1. If $x^1 = \begin{bmatrix} 1 \\ -2 \end{bmatrix}$, $x^2 = \begin{bmatrix} 2 \\ 5 \end{bmatrix}$, $y = \begin{bmatrix} 1 \\ 5 \end{bmatrix}$, find c_1 and c_2 such that $y = c_1 x^1 + c_2 x^2$.
2. Determine the dimension of the vector space of the set of vectors

$$x^T = [1 \quad 3 \quad 0], \quad y^T = [2 \quad -5 \quad 0], \quad z^T = [-1 \quad 7 \quad 0].$$

3. Explain why we cannot write $y = \begin{bmatrix} 1 \\ 2 \\ 5 \end{bmatrix}$ in terms of the two non-zero orthogonal vectors $\begin{bmatrix} 1 \\ -1 \\ 0 \end{bmatrix}$ and $\begin{bmatrix} 4 \\ 4 \\ 0 \end{bmatrix}$.
4. Show that $\sum_{i=1}^{n} \langle x^i, y \rangle x^i = y$, if $\langle x^i, x^j \rangle = \delta_{ij}$, with $\delta_{ij} = 1 (i = j)$, $\delta_{ij} = 0 \, (i \neq j)$.
5. If $\langle x^i, x^j \rangle = 1$, $(i = j)$, and $\langle x^i, x^j \rangle = 0$, $(i \neq j)$, show that

$$\langle y, y \rangle = \sum_{i=1}^{n} |c_i|^2, \quad \text{if } y = \sum_{i=1}^{n} c_i x^i$$

6. If y is orthogonal on $x^1, \ldots, x^m$, show that y is orthogonal on any vector which is a linear combination of $x^1, \ldots, x^m$.
7. If $\langle x^i, x^j \rangle = 1$, $(i = j)$, and $\langle x^i, x^j \rangle = 0$ $(i \neq j)$ show that the matrix $A_{n,n}$ whose columns are $x^1, \ldots, x^n$ is unitary.
8. Expand $y = \begin{bmatrix} i \\ 3 \end{bmatrix}$, $(i^2 = -1)$, in terms of $x^1 = \begin{bmatrix} 1 \\ 2 \end{bmatrix}$ and $x^2 = \begin{bmatrix} -2 \\ 1 \end{bmatrix}$
9. Determine the rank of the vectors $\begin{bmatrix} 1 \\ 3 \\ 2 \end{bmatrix}, \begin{bmatrix} 1 \\ 2 \\ 5 \end{bmatrix}, \begin{bmatrix} 5 \\ -2 \\ 0 \end{bmatrix}, \begin{bmatrix} 1 \\ 0 \\ 1 \end{bmatrix}, \begin{bmatrix} -1 \\ 4 \\ 7 \end{bmatrix}$.
10. Determine the rank of the vectors $\begin{bmatrix} 1 \\ i \\ 0 \end{bmatrix}, \begin{bmatrix} 2+i \\ -2 \\ 1 \end{bmatrix}, \begin{bmatrix} 4+3i \\ -5 \\ 2 \end{bmatrix}$ $(i^2 = -1)$.
11. Under what condition will the rank of the vectors $\begin{bmatrix} 1 \\ 0 \\ 0 \end{bmatrix} \begin{bmatrix} 0 \\ r-2 \\ 2 \end{bmatrix} \begin{bmatrix} 0 \\ k-1 \\ k+2 \end{bmatrix}$ $\begin{bmatrix} 0 \\ 0 \\ 3 \end{bmatrix}$ be less than three?
12. Prove that A^*A is diagonal if and only if the columns of A are mutually orthogonal, and that $AA^* = I$ if and only if the rows of A are mutually orthogonal unit vectors (A need not be square).
13. Show that, if x, y and z are mutually orthogonal vectors,

$$\| x + y + z \|_2 = (\| x \|_2^2 + \| y \|_2^2 + \| z \|_2^2)^{1/2}.$$

14. If $x^1, \ldots, x^m$ are linearly independent, show that $\bar{x}^1, \ldots, \bar{x}^m$ are also linearly independent.
15. Let L be the linear space of finite trigonometric sums

$$x = c_1 \sin t + c_2 \sin 2t + \cdots + c_k \sin kt \quad (0 \leqslant t \leqslant 2\Pi)$$

where $c_1, c_2, \ldots$ are real coefficients. Show that the vectors $\sin t, \ldots, \sin kt$ constitute a basis for L. Hence show how to obtain $c_1, c_2, \ldots$.
16. If $\langle x^i, y^j \rangle = 1$, $(i = j)$, and $\langle x^i, y^j \rangle = 0$, $(i \neq j)$, show that

$\langle v, v \rangle = \sum_{i=1}^{n} \alpha_i \bar{c}_i = \sum_{i=1}^{n} \bar{\alpha}_i c_i$, if $v = \sum_{i=1}^{n} c_i x^i$ and $v = \sum_{i=1}^{n} \alpha_i y^i$, which gives a relation between c_i and α_i; hence show that $\sum_{i=1}^{n} \alpha_i \bar{c}_i$ and $\sum_{i=1}^{n} \bar{\alpha}_i c_i$ are real, for both are equal to $\langle v, v \rangle$, which is real. Finally show that $\sum_{i=1}^{n} \bar{\alpha}_i c_i = \langle \alpha, G_y \alpha \rangle$, where $\alpha^T = [\alpha_1 \alpha_2 \ldots \alpha_n]$ and G_y is the Grammian matrix for the vectors y^i.
17. If $x^1, x^2, \ldots, x^m$ are linearly independent as well as $y^1, y^2, \ldots, y^s$, and $\langle x^i, y^j \rangle = 0$ for all i, j, show that $x^1, \ldots, x^m, y^1, \ldots, y^s$ are linearly independent.
18. If $x^1, x^2, \ldots, x^n \in \mathbb{R}^n$ are linearly independent and are all orthogonal on a vector v, show that $v = 0$.

19. If $u^1, u^2, \ldots, u^m$ are a set of linearly independent vectors, show that the vectors $v^1, v^2, \ldots, v^m$ where $v^i = \sum_k \alpha_k^i u^k$, with α_k^i arbitrary and not zero, are also a set of linearly independent vectors. Show also that the set $v^1, v^2, \ldots, v^m, v^{m+1}$ are linearly dependent.

3.6. Rank of a matrix

Let $A_{m,n}$ be a matrix which contains m rows and n columns. Assume without loss of generality that $m \leqslant n$. A can be seen as consisting of m vectors in an n-dimensional Euclidean space, the rank of which will be denoted by ρ_r (row rank). Obviously

$$\rho_r \leqslant m$$

Also A can be seen as consisting of n vectors in an m-dimensional Euclidean space, the rank of which will be denoted by ρ_c (column rank). Obviously

$$\rho_c \leqslant m$$

The following theorem establishes the relation between the row rank and the column rank of a matrix.

THEOREM 1. In a matrix A, the number of linearly independent row vectors is equal to the number of linearly independent column vectors. In other words row rank ρ_r = column rank ρ_c = rank of A denoted by ρ.

Proof. Let the number of linearly independent column vectors in $A_{m,n}$ be ρ_c and the number of linearly independent row vectors be ρ_r. Select ρ_c linearly independent columns of A and put them in a matrix B_1 whose dimension is (m, ρ_c). As a result of Theorem 5 of Section 3.1 every column of A can be written as a linear combination of these independent vectors. If $x^1, x^2, \ldots, x^n$ are the columns of A, then

$$x^1 = B_1 v^1$$
$$x^2 = B_1 v^2$$
$$\vdots$$
$$x^n = B_1 v^n$$

where the components of v^i are the coordinates of x^i w.r.t. the basis consisting of the column vectors of B_1. Hence

$$A = [B_1 v^1 \; B_1 v^2 \ldots B_1 v^n] = B_1 [v^1 v^2 \ldots v^n] = B_1 B_2.$$

This means that the row vectors of A are linear combinations of the row vectors of B_2. It follows that the rank of the system of row vectors of A cannot exceed the number of row vectors in B_2. This means that

$$\rho_r \leqslant \rho_c.$$

If we consider the transpose of A we obtain

$$\rho_c \leqslant \rho_r,$$

from which we conclude that

$$\rho_r = \rho_c$$

Example

$$A_{4,2} = \begin{bmatrix} 1 & -1 \\ 0 & 2 \\ 3 & -1 \\ 1 & 0 \end{bmatrix}$$

The matrix A has two columns in $\mathbb{R}^4$; hence $\rho_c \leqslant 2$. One can use the test of independence to show that $\rho_c = 2$. A also has four vectors in $\mathbb{R}^2$, hence $\rho_r \leqslant 2$. Similarly it can be shown that $\rho_r = 2$. Hence $\rho_r = \rho_c = 2$. In this example the rank is obtained easily because it is equal to the number of columns, and so calculating it is equivalent to making sure that the two columns are linearly independent, in other words by using the test of independence. However in some examples we find that the rank is less than both the number of rows and the number of columns; then we need methods for calculating the rank.

THEOREM 2. Let A be a matrix of order (m, n). Suppose that A has a sub-matrix S of order (r, r) with $|S| \neq 0$. And suppose that every sub-matrix T of order $(r+1, r+1)$ of which S is a sub-matrix has $|T| = 0$. Then $\rho(A) = r$.

Before proceeding with the proof, let us explain the theorem by an example. Let

$$A = \begin{bmatrix} 0 & 1 & 0 & 1 & 2 & 0 & 3 \\ 0 & 2 & 0 & 2 & 4 & 0 & 6 \\ 0 & 1 & 0 & 2 & 4 & -1 & 4 \end{bmatrix}$$

The submatrix

$$S = \begin{bmatrix} 1 & 1 \\ 1 & 2 \end{bmatrix}$$

has $|S| \neq 0$, but every $(3, 3)$ sub-matrix T including S has $|T| = 0$, e.g.

$$\begin{vmatrix} 1 & 1 & 2 \\ 2 & 2 & 4 \\ 1 & 2 & 4 \end{vmatrix} = 0, \quad \begin{vmatrix} 1 & 1 & 0 \\ 2 & 2 & 0 \\ 1 & 2 & -1 \end{vmatrix} = 0, \quad \begin{vmatrix} 0 & 1 & 1 \\ 0 & 2 & 2 \\ 0 & 1 & 2 \end{vmatrix} = 0.$$

Hence from the theorem we conclude that $\rho(A) = 2$. Now we proceed with the proof.

Proof. Let S be composed of the linearly independent row vectors $x^1, \ldots, x^r$ in the following manner:

$$S = \begin{bmatrix} x^1 \\ x^2 \\ \vdots \\ x^r \end{bmatrix}$$

Construct a sub-matrix S' as follows

$$S' = [S \,\vdots\, v] = \begin{bmatrix} x'^1 \\ \vdots \\ x'^r \end{bmatrix}$$

where v is any column vector. Now S' has rank $= r$, as the reader can verify. And as T can be written without loss of generality in the following form

$$T = \begin{bmatrix} S' \\ \hdashline u \end{bmatrix};$$

then $|T| = 0$ implies that the row vectors of T are linearly dependent, i.e.

$$c_1 x'^1 + c_2 x'^2 + \cdots + c_r x'^r + c_{r+1} u = 0$$

where

$$c_{r+1} \neq 0;$$

otherwise all the other coefficients c_k will be zero since $x'^1, x'^2, \ldots, x'^r$ are linearly independent, hence

$$u = \alpha_1 x'^1 + \alpha_2 x'^2 + \cdots + \alpha_r x'^r$$

where $\alpha_1, \alpha_2, \ldots, \alpha_r$ are scalars.

Applying the same procedure to all sub-matrices of order $(r + 1, r + 1)$ in which S is a submatrix; and use Theorem 7 in Section 3.1 to complete the proof.

One direct application of this theorem is when A is Hermitian of rank r; then at least one principal minor of order r is not zero. The proof of this corollary is left as an exercise for the reader.

3.7. Elementary row and column operations

There are sets of elementary row and column operations which can be executed on any matrix A to reduce it to an echelon form. These operations can be achieved by pre-multiplying A by a set of matrices (R_1, R_2, R_3) for rows or by post-multiplying A by a set of matrices (P_1, P_2, P_3) for columns. The elementary row operations are of three types:

R_1, responsible for interchanging any two rows

R_2, responsible for multiplying any row by a non-zero scalar
R_3, responsible for adding to any row any other row multiplied by a non-zero scalar.

R_1, R_2 and R_3 are all generated from the unit matrix I, by applying on the latter the same change which we require for the matrix A. For example if we want to interchange the first and the third rows of A, we multiply A by R_1, which is a unit matrix whose first and third rows are interchanged.

Example. Let

$$A = \begin{bmatrix} 1 & 2 & 3 & 0 \\ -2 & 1 & 0 & 4 \\ 5 & -2 & 3 & 1 \end{bmatrix}, \quad A' = \begin{bmatrix} 5 & -2 & 3 & 1 \\ -2 & 1 & 0 & 4 \\ 1 & 2 & 3 & 0 \end{bmatrix}$$

Hence

$$R_1 A = A' \text{ where } R_1 = \begin{bmatrix} 0 & 0 & 1 \\ 0 & 1 & 0 \\ 1 & 0 & 0 \end{bmatrix}$$

The reader is asked to prove that elementary row operations do not change the rank of the matrix on which they operate; that R_1, R_2 and R_3 are nonsingular matrices; and that $|R_1| = -1$, $|R_3| = 1$. What is the value of $|R_2|$?

THEOREM 1. Any matrix can be reduced to an echelon form by a series of elementary row operations.

The proof is by construction and is best illustrated on an example. Let

$$A = \begin{bmatrix} 5 & 2 & 3 & -4 \\ 2 & 1 & 0 & 2 \\ -3 & -1 & -3 & 6 \end{bmatrix}$$

Step 1: divide the first row by 5, giving

$$A_1 = \begin{bmatrix} 1 & 2/5 & 3/5 & -4/5 \\ 2 & 1 & 0 & 2 \\ -3 & -1 & -3 & 6 \end{bmatrix} \quad \text{with } R_2^1 = \begin{bmatrix} 1/5 & 0 & 0 \\ 0 & 1 & 0 \\ 0 & 0 & 1 \end{bmatrix}$$

Step 2: multiply the first row by 2 and subtract it from the second row, then multiply it by 3 and add it to the third row, giving

$$A_2 = \begin{bmatrix} 1 & 2/5 & 3/5 & -4/5 \\ 0 & 1/5 & -6/5 & 18/5 \\ 0 & 1/5 & -6/5 & 18/5 \end{bmatrix} \quad \text{with } R_3^1 = \begin{bmatrix} 1 & 0 & 0 \\ -2 & 1 & 0 \\ 3 & 0 & 1 \end{bmatrix}$$

Step 3: multiply the second row by 5, giving

$$A_3 = \begin{bmatrix} 1 & 2/5 & 3/5 & -4/5 \\ 0 & 1 & -6 & 18 \\ 0 & 1/5 & -6/5 & 18/5 \end{bmatrix} \quad \text{with } R_2^2 = \begin{bmatrix} 1 & 0 & 0 \\ 0 & 5 & 0 \\ 0 & 0 & 1 \end{bmatrix}$$

Step 4: multiply the second row by 1/5 and subtract it from the third row, giving

$$A_4 = \begin{bmatrix} 1 & 2/5 & 3/5 & -4/5 \\ 0 & 1 & -6 & 18 \\ 0 & 0 & 0 & 0 \end{bmatrix} \quad \text{with } R_3^2 = \begin{bmatrix} 1 & 0 & 0 \\ 0 & 1 & 0 \\ 0 & -1/5 & 1 \end{bmatrix}$$

Step 5: multiply the second row by 2/5 and subtract it from the first row, giving

$$A_5 = \begin{bmatrix} 1 & 0 & 3 & -8 \\ 0 & 1 & -6 & 18 \\ 0 & 0 & 0 & 0 \end{bmatrix} \quad \text{with } R_3^3 = \begin{bmatrix} 1 & -2/5 & 0 \\ 0 & 1 & 0 \\ 0 & 0 & 1 \end{bmatrix}$$

Hence

$$A_5 \text{ (in echelon form)} = R_3^3 R_3^2 R_2^2 R_3^1 R_2^1 A$$

The reader will notice that the rank of A_5 is 2; hence the rank of A is also 2. Thus the above technique of elementary row operations applied on any matrix to transform it to an echelon form can be a very easy and systematic method for calculating the rank of a matrix. The rank of A is equal to the number of non-zero rows in its reduced echelon form.

THEOREM 2. Any nonsingular matrix can be transformed into a unit matrix by elementary row operations, i.e. a nonsingular matrix possesses a unit matrix as its echelon form.

The proof is given by construction and is best demonstrated by an example. Let

$$A = \begin{bmatrix} 10 & 2 & 4 \\ 1 & 3 & 2 \\ -2 & 4 & 1 \end{bmatrix}$$

Step 1: we either divide the first row by 10 or, better, interchange the first row with the second, giving

$$A_1 = \begin{bmatrix} 1 & 3 & 2 \\ 10 & 2 & 4 \\ -2 & 4 & 1 \end{bmatrix} \quad \text{with } R_1^1 = \begin{bmatrix} 0 & 1 & 0 \\ 1 & 0 & 0 \\ 0 & 0 & 1 \end{bmatrix}$$

Step 2: multiply the first row by 10 and subtract it from the second row, and then multiply the first row by -2 and subtract it from the third row, giving

$$A_2 = \begin{bmatrix} 1 & 3 & 2 \\ 0 & -28 & -16 \\ 0 & 10 & 5 \end{bmatrix} \quad \text{with } R_3^1 = \begin{bmatrix} 1 & 0 & 0 \\ -10 & 1 & 0 \\ 2 & 0 & 1 \end{bmatrix}$$

Step 3: multiply the second row by $-1/28$, giving

$$A_3 = \begin{bmatrix} 1 & 3 & 2 \\ 0 & 1 & 4/7 \\ 0 & 10 & 5 \end{bmatrix} \quad \text{with } R_2^1 = \begin{bmatrix} 1 & 0 & 0 \\ 0 & -1/28 & 0 \\ 0 & 0 & 1 \end{bmatrix}$$

Step 4: multiply the second row by 10 and subtract it from the third, giving

$$A_4 = \begin{bmatrix} 1 & 3 & 2 \\ 0 & 1 & 4/7 \\ 0 & 0 & -5/7 \end{bmatrix} \quad \text{with } R_3^2 = \begin{bmatrix} 1 & 0 & 0 \\ 0 & 1 & 0 \\ 0 & -10 & 1 \end{bmatrix}$$

Step 5: multiply the third row by $-7/5$, giving

$$A_5 = \begin{bmatrix} 1 & 3 & 2 \\ 0 & 1 & 4/7 \\ 0 & 0 & 1 \end{bmatrix} \quad \text{with } R_2^2 = \begin{bmatrix} 1 & 0 & 0 \\ 0 & 1 & 0 \\ 0 & 0 & -7/5 \end{bmatrix}$$

Step 6: multiply the third row by 4/7 and subtract it from the second, giving

$$A_6 = \begin{bmatrix} 1 & 3 & 2 \\ 0 & 1 & 0 \\ 0 & 0 & 1 \end{bmatrix} \quad \text{with } R_3^3 = \begin{bmatrix} 1 & 0 & 0 \\ 0 & 1 & -4/7 \\ 0 & 0 & 1 \end{bmatrix}$$

Step 7: multiply the third row by 2, the second row by 3 and subtract each of them from the first, giving

$$A_7 = \begin{bmatrix} 1 & 0 & 0 \\ 0 & 1 & 0 \\ 0 & 0 & 1 \end{bmatrix} \quad \text{with } R_3^4 = \begin{bmatrix} 1 & -3 & -2 \\ 0 & 1 & 0 \\ 0 & 0 & 1 \end{bmatrix}$$

Hence

$$A_7 = I = R_3^4 R_3^3 R_2^2 R_3^2 R_2^1 R_3^1 R_1^1 \, A$$

What has been said about elementary row operations can be extended to elementary column operations. The latter are similarly of three types:

P_1, responsible for interchanging any two columns
P_2, responsible for multiplying any column by a non-zero scalar
P_3, responsible for adding to any column any other column multiplied by a non-zero scalar.

P_1, P_2 and P_3 are all generated from the unit matrix I exactly like the elementary row operations except that the operations are done on the columns instead of the rows.

THEOREM 3. Any matrix A can be reduced to the transpose of an echelon form by a series of elementary column operations. Moreover, if A is nonsingular it will be transformed into a unit matrix.

As an application of the above theorems we can add new ones, whose proof is left as an exercise for the reader.

THEOREM 4. If A is of order (m, n), $m > n$ and $\rho(A) = n$, then A can be transformed, by elementary row operations only, into the form

$$\begin{bmatrix} I_{n,n} \\ \hdashline 0 \end{bmatrix}$$

THEOREM 5. If A is of order (m, n), $m < n$ and $\rho(A) = m$, then A can be transformed, by elementary column operations only, into the form

$$[I_{m,m} \;\vdots\; 0]$$

THEOREM 6. If A is of order (m, n), $\rho(A) < m$, $\rho(A) < n$, then A can be transformed by a series of both elementary row and column operations into the form

$$\begin{bmatrix} I_{\rho,\rho} & \vdots & 0 \\ \hdashline 0 & \vdots & 0 \end{bmatrix}$$

The last three forms shown are known together as the *normal form* of a matrix A. These forms are sometimes obtained when one needs to calculate the rank of a matrix. As is apparent, obtaining the normal forms is a simple way to calculate $\rho(A)$.

3.8. Rank of sum and product of two matrices

In this section we prove two useful theorems relating the rank of two matrices to the rank of their sum or product. We begin by the rank of the sum of two matrices and establish the following theorem

THEOREM 1.

$$\rho(A + B) \leqslant \rho(A) + \rho(B).$$

Proof. Let $C = A + B$ and we proceed to show that $\rho(C) \leqslant \rho(A) + \rho(B)$. Apply on A a set of elementary row operations until it is transformed into an echelon form like the following:

$$C' = \begin{matrix} \rho(A)\{ \\ \\ \end{matrix} \begin{bmatrix} A_1 \\ \hdashline 0 \end{bmatrix} + \begin{bmatrix} B_1 \\ \hdashline B_2 \end{bmatrix}$$

where C' has the same rank as C. Next operate on B_2 by a set of elementary row operations until B_2 is transformed also into an echelon form like the following:

$$C'' = \begin{matrix} \rho(A)\{ \\ \\ \\ \end{matrix} \begin{bmatrix} A_1 \\ \hdashline \\ 0 \end{bmatrix} + \begin{bmatrix} B_1 \\ \hdashline 0 \\ \hdashline B_3 \end{bmatrix} \begin{matrix} \\ \\ \}\rho(B_2) \end{matrix}$$

By a series of elementary column operations B_3 can be put in the following form:

$$C''' = \begin{array}{l}\rho(A)\{\\ \\ \\ \end{array}\left[\begin{array}{c:c} A_2 & A_3 \\ \hdashline \multicolumn{2}{c}{0} \\ \multicolumn{2}{c}{} \end{array}\right] + \left[\begin{array}{c:c} B_4 & B_5 \\ \hdashline 0 & 0 \\ \hdashline 0 & I \end{array}\right]\}\rho(B_2)$$

Now adding we obtain

$$C''' = \begin{array}{l}\rho(A)\{\\ \\ \\ \end{array}\left[\begin{array}{c:c} A_2 + B_4 & A_3 + B_5 \\ \hdashline 0 & 0 \\ \hdashline 0 & I \end{array}\right]\}\rho(B_2)$$

Finally, by using a series of elementary row operations, the above matrix takes the form:

$$C'''' = \begin{array}{l}\rho(A)\{\\ \\ \\ \end{array}\left[\begin{array}{c:c} A_2 + B_4 & 0 \\ \hdashline 0 & 0 \\ \hdashline 0 & I \end{array}\right]\}\rho(B_2)$$

Therefore

$$\begin{aligned}\rho(C'''') = \rho(C) = \rho(B_2) + \rho(A_2 + B_4) \\ \leqslant \rho(B_2) + \rho(A) \\ \leqslant \rho(B) + \rho(A)\end{aligned}$$

and the proof is complete.

Example. Let

$$A = \begin{bmatrix} 1 & 2 & -1 \\ 0 & 1 & 2 \\ 1 & 2 & 1 \end{bmatrix}, \quad B = \begin{bmatrix} 1 & 0 & 1 \\ 2 & -1 & 0 \\ 3 & -1 & 1 \end{bmatrix}$$

Then $\rho(A + B) = 3$; however $\rho(A) = 3$; $\rho(B) = 2$.

THEOREM 2. $\rho(AB) \geqslant \rho(A) + \rho(B) - n$, where n is the number of rows of B.

Proof. Post-multiplying B by a set of elementary column operations, we obtain from Theorem 3 in Section 3.7:

$$BP = [u^1 \vdots u^2 \vdots \cdots \vdots u^{\rho(B)} \vdots 0]$$

where the vectors $u^1, u^2, \ldots, u^{\rho(B)}$ are linearly independent. Now pre-multiplying BP by A we obtain

$$ABP = [Au^1 \vdots Au^2 \vdots \cdots \vdots Au^{\rho(B)} \vdots 0]$$

of which we seek to determine the rank.

Now consider the equation

$$Ax^i = 0$$

The number of linearly independent vectors x^i which satisfy the above equation are equal to $n - \rho(A)$, as the reader can verify. From this we conclude that the number of linearly dependent column vectors in ABP are $\rho(B) \quad \rho(ABP) = \rho(B) - \rho(AB)$ must not exceed $n - \rho(A)$. i.e.

$$\rho(B) - \rho(AB) \leqslant n - \rho(A)$$

and the proof is complete.

Example

$$A = \begin{bmatrix} 1 & 2 & 1 \\ 0 & 2 & 1 \end{bmatrix}, \quad B = \begin{bmatrix} 2 & -4 \\ -1 & 2 \\ 2 & 1 \end{bmatrix}, \quad AB = \begin{bmatrix} 2 & 1 \\ 0 & 5 \end{bmatrix}$$

$\rho(AB) = 2$, $\rho(A) = 2$, $\rho(B) = 2$, $n = 3$.

Exercises 3.8

1. Transform into normal form the matrices

$$\begin{bmatrix} 1 & 2 \\ -1 & 3 \\ 0 & 1 \\ 3 & -2 \end{bmatrix}, \quad \begin{bmatrix} 2 & 3 & -1 & 4 \\ 2 & -1 & 0 & 2 \end{bmatrix}, \quad \begin{bmatrix} 5 & 2 & 3 & -4 \\ 2 & 1 & 0 & 2 \\ -3 & -1 & -3 & 6 \\ -3 & -1 & -3 & 6 \end{bmatrix}$$

and find for each case the required elementary row and column operations R and P.

2. Show that rank $\begin{bmatrix} A & 0 \\ 0 & B \end{bmatrix}$ = rank A + rank B.

3. If either A or B is nonsingular, show that

$$\text{rank} \begin{bmatrix} A & 0 \\ C & B \end{bmatrix} = \text{rank } A + \text{rank } B.$$

4. Show that $\rho(AB) \leqslant \min\{\rho(A), \rho(B)\}$.
5. If A is of order (m, n) and B of order (n, m) and if $m > n$, show that AB is singular. Give a separate proof from that of Exercise 2.1.18.
6. Show that any nonsingular matrix A can be factored into the product of elementary row operations, i.e. $A = R_1 R_2 \ldots R_q$. *Hint*: define a matrix A^{-1} such that $A^{-1}A = I$ and transform A^{-1} into a unit matrix I by a series of elementary row operations.
7. Two matrices A and B are said to be equivalent if they have the same order and rank. Show that there exist nonsingular matrices C and D such that $CAD = B$, for any A and B.
8. Show that $\rho(\alpha A) = \rho(A)$, where α is a non-zero scalar.
9. If $\begin{bmatrix} A & B \\ C & I \end{bmatrix}$ is nonsingular with $AC^T + B = 0$, show that A is nonsingular.

10. If $\begin{bmatrix} A \\ \hdashline B \end{bmatrix}$ is nonsingular, with $AB^T = 0$, show that it can be transformed by elementary operations into the form $\left[\begin{array}{c:c} I & D \\ \hdashline -D^T & I \end{array}\right]$.

11. Under what condition will the matrix $\begin{bmatrix} A & B \\ C & D \end{bmatrix}$ be nonsingular if A is nonsingular?

12. If A is Hermitian, does there exist an R such that R^*AR = normal form?

13. Show that $\rho(AA^*) = \rho(A^*A) = \rho(A)$.

14. Show that if A is of order (m, n), and $m > n$, then AA^* is singular.

15. Show that if A is of order (m, n) and $m < n$ and $\rho(A) = m$, then AA^* is nonsingular.

16. If $x = Ty$, with T nonsingular, deduce a technique for making $y = T'x$. *Hint*: the operations can be done one at a time. For example we start by interchanging x_1 and y_1 such that the equations become as shown

$$\begin{bmatrix} y_1 \\ x_2 \\ \vdots \\ x_n \end{bmatrix} = \begin{bmatrix} \\ & T_1 & \\ \\ \end{bmatrix} \begin{bmatrix} x_1 \\ y_2 \\ \vdots \\ y_n \end{bmatrix}$$

This method is due to Gauss–Jordan; see also Stoer and Bulirsch (1980), p. 169.

17. If AB is nonsingular, with A and B being square, show that A and B are nonsingular. Give a different proof from that of Exercise 2.1.12.

18. Show that the product of nonsingular matrices is also nonsingular.

19. If A is nonsingular, show that $\rho(AB) = \rho(B)$.

20. Show that $\rho(A - B) \geqslant \rho(A) - \rho(B)$.

21. Show that $\rho(AB) + \rho(BC) \leqslant \rho(B) + \rho(ABC)$, whenever ABC exists.

22. Show that $\rho(A + B + C) \leqslant \rho(A) + \rho(B) + \rho(C)$.

23. If A is of order (m, n), B of order (n, p) and $AB = 0$, show that $\rho(A) + \rho(B) \leqslant n$.

24. Given an arbitrary square matrix A of rank $\rho(A)$, the set of all vectors x which transform into the zero vector, i.e. which satisfies the equation $Ax = 0$, is called the *null-space* of A. If A has order n, the null space must then have dimension $n - \rho(A)$, which is the number of independent solutions of $Ax = 0$. The dimension of the null-space is called the *nullity* of A and is denoted by $N(A)$, so that $N(A) = n - \rho(A)$. Show that

(i) $N(AB) \geqslant \max\{N(A), N(B)\}$
(ii) $N(AB) \leqslant N(A) + N(B)$
(iii) $N(A + B) \geqslant N(A) + N(B) - n$

25. Show how to use the elementary operations to compute the determinant of a square matrix A. *Hint*: transform A into upper triangular form as follows:

$$\begin{bmatrix} a_{11} & a_{12} & \cdots & a_{1n} \\ & a_{22} & a_{23} & \vdots \\ 0 & & \ddots & a_{n-1,n} \\ & & & a_{nn} \end{bmatrix}$$

This method is the one used on a computer to calculate $|A|$ since it is much easier and faster than calculating cofactors. Also if A is singular, how can one realize it directly?

26. If from a square matrix A of order n and rank $\rho(A)$, a sub-matrix B consisting of s rows (columns) of A is selected, show that $\rho(B) \geqslant \rho(A) + s - n$. *Hint*: $N(A) \geqslant N(B)$.

27. Show how the above exercise can help in proving that $\rho(AB) \geqslant \rho(A) + \rho(B) - n$ even if A and B are not square, and where n is the number of rows of B. *Hint*: consider $\rho(RAPB) = \rho(AB)$, if $RAP = \begin{bmatrix} I_{\rho(A)} & 0 \\ 0 & 0 \end{bmatrix}$

28. If $|A| \neq 0$, show that $Ax = 0$ implies $x = 0$. Give a proof using the rank of the product.

3.9. Solution of a system of linear simultaneous equations

Let the equations be written in the form:

$$\begin{aligned} a_{11}x_1 + a_{12}x_2 + \cdots + a_{1n}x_n &= b_1 \\ a_{21}x_1 + a_{22}x_2 + \cdots + a_{2n}x_n &= b_2 \\ a_{m1}x_1 + a_{m2}x_2 + \cdots + a_{mn}x_n &= b_m \end{aligned}$$

where it is usually required to solve for $x_1, \ldots, x_n$, when the as and bs have known values. The above set of equations can be written in the abbreviated form

$$Ax = b$$

where the dimensions of A, x and b are respectively (m, n), $(n, 1)$ and $(m, 1)$. Therefore the problem that we pose is to find a vector x which satisfies the above equation if A and b are known.

For this we face three situations:

(a) The number of equations is equal to the number of unknowns, i.e. $m = n$.
(b) The number of equations is less than the number of unknowns i.e. $m < n$.
(c) The number of equations is greater than the number of unknowns i.e. $m > n$.

One may suppose that the relation between the number of equations and the number of unknowns is the one which decides in general whether the equations have either a unique solution for x, more than one solution for x or none at all. Such a supposition can be easily destroyed through a simple counter-example. Consider the three sets of equations where $m = n = 2$.

$$\begin{bmatrix} 1 & 2 \\ -1 & 1 \end{bmatrix}\begin{bmatrix} x_1 \\ x_2 \end{bmatrix} = \begin{bmatrix} 1 \\ 1 \end{bmatrix}, \quad \begin{bmatrix} 1 & 2 \\ 2 & 4 \end{bmatrix}\begin{bmatrix} x_1 \\ x_2 \end{bmatrix} = \begin{bmatrix} -1 \\ -2 \end{bmatrix},$$

$$\begin{bmatrix} 1 & 2 \\ 2 & 4 \end{bmatrix}\begin{bmatrix} x_1 \\ x_2 \end{bmatrix} = \begin{bmatrix} 1 \\ 0 \end{bmatrix}$$

One can realize through simple manipulations that the first set has a unique solution for x_1, x_2; that the second set has more than one solution for x_1, x_2; and that the third set has no solution for x_1x_2, i.e. it is inconsistent. Geometrically the first set represents two intersecting lines, the second two superimposed lines and the third two parallel lines. Therefore the above classification is useless.

Moreover to solve equations with $m < n$, we may need to assign arbitrary values to more than $n - m$ of the unknowns. That can be answered only when we compute the rank of A; for there may be redundant equations. Not only that, but the vector b should also be checked to see whether redundancy is also exported to it, as in the second example above. In short, what determines the behaviour of x is the rank of A, together with its relation to b. The above classification is therefore altered to the following three new ones:

(a) *A is nonsingular*
If $b = 0$, then $x = 0$ (see Exercise 2.1.14), which is termed the trivial solution for the equations.

If $b \neq 0$ we consider different numerical methods to solve for x. They are usually of two types, direct and iterative. Direct methods are the ones which yield exact solutions in a finite number of steps of arithmetic operations if round-off errors are absent. Unless the number of equations is small and the element values of A are well equilibrated together, in practice direct methods do not lead to exact solutions since they are executed on the computer, where we usually face round-off errors as well as errors arising from instability or loss of significance in the figures. One of the best direct methods is the Gauss elimination, but even with this there is a choice of alternatives which differ in duration, computational efficiency and accuracy. On the other hand, iterative methods have the advantage of simplicity, uniformity and accuracy. They start by assigning an initial guess to the vector x and by applying a suitably chosen iterative algorithm; one hopes that the process converges to the required solution within a prescribed accuracy. Sometimes we can only work with iterative methods as in large systems when most of the time the matrix A is *sparse* (not dense), i.e. having very few non-zero elements. Such a situation appears frequently when solving differential equations by the finite-difference method. The only disadvantage of iterative methods is that the matrix A of coefficients should have a special property or be rearranged to possess this property in order to be qualified to enter the algorithm. Otherwise convergence may be very slow or non-existing at all.

Of the direct methods that are available we mention Cramer's rule, Gauss elimination and Cholesky decomposition in order of efficiency. Cramer's rule is the most troublesome since it involves the evaluation of determinants. To explain the method, let the equations be written in the form:

$$\sum_{k=1}^{n} a_{1k}x_k = b_1$$

$$\vdots$$

$$\sum_{k=1}^{n} a_{nk}x_k = b_n$$

Multiplying the first equation by the cofactor C_{11}, the second by C_{21}, etc., up to the last one by C_{n1}, and adding, we obtain

$$x_1 \sum_{k=1}^{n} a_{k1}C_{k1} + x_2 \sum_{k=1}^{n} a_{k2}C_{k1} + \cdots + x_n \sum_{k=1}^{n} a_{kn}C_{k1} = \sum_{k=1}^{n} b_k C_{k1}$$

where all the terms from the second to the last term in the left-hand side of the equation are equal to zero from Property 8 of determinants in Section 2.1; hence

$$x_1 = \frac{\sum_{k=1}^{n} b_k C_{k1}}{\sum_{k=1}^{n} a_{k1} C_{k1}} = \frac{\begin{vmatrix} b_1 & a_{12} & \cdots & a_{1n} \\ b_2 & a_{22} & & \\ & & & \\ b_n & a_{n2} & & a_{nn} \end{vmatrix}}{|A|}$$

in other words x_1 is obtained by dividing the determinant (of the matrix obtained from A by replacing its first column by the vector b) by the determinant of A; in general

$$x_k = \frac{\begin{vmatrix} a_{11} & \cdots & a_{1,k-1} b_1 & a_{1,k+1} & \cdots & a_{1n} \\ a_{21} & \cdots & a_{2,k-1} b_2 & a_{2,k+1} & \cdots & a_{2n} \\ & & & & & \\ a_{n1} & \cdots & a_{n,k-1} b_n & a_{n,k+1} & \cdots & a_{nn} \end{vmatrix}}{|A|}$$

Example. Solve the equations

$$x_1 + 2x_2 = 3, \quad 2x_1 - 2x_2 = 24$$

$$x_1 = \frac{\begin{vmatrix} 3 & 2 \\ 24 & -2 \end{vmatrix}}{\begin{vmatrix} 1 & 2 \\ 2 & -2 \end{vmatrix}} = 9, \quad x_2 = \frac{\begin{vmatrix} 1 & 3 \\ 2 & 24 \end{vmatrix}}{\begin{vmatrix} 1 & 2 \\ 2 & -2 \end{vmatrix}} = -3$$

It is to be noticed that in practice the solution provided by Cramer's rule is almost impossible to attain since it involves the evaluation of $(n + 1)$ determinants of order n, where each requires a huge amount of arithmetical operations. Young and Gregory (1973), p. 790, report that for $n = 20$, the number of arithmetical calculations would rise to nearly 16×10^{19}. Surprisingly enough, a computer performing 2 million operations per second would spend on the problem 2 million years.

A far better algorithm is the one due to Gauss which relies upon operating on both parts of the equations by a series of elementary row operations until the matrix A is transformed into a unit matrix as follows:

$$RAx = Rb$$

such that

$$RA = I$$

Hence

$$x = Rb = b'$$

Example. Solve the equations

$$x_1 + 2x_2 + 3x_3 = 1, \quad -x_1 + 2x_3 = 2, \quad 3x_1 + x_2 + 2x_3 = 1$$

The equations are put in the matrix form

$$\begin{bmatrix} 1 & 2 & 3 \\ -1 & 0 & 2 \\ 3 & 1 & 2 \end{bmatrix} \begin{bmatrix} x_1 \\ x_2 \\ x_3 \end{bmatrix} = \begin{bmatrix} 1 \\ 2 \\ 1 \end{bmatrix}$$

By applying a set of elementary row operations we obtain successively:

$$\begin{bmatrix} 1 & 2 & 3 \\ 0 & 2 & 5 \\ 0 & -5 & -7 \end{bmatrix} \begin{bmatrix} x_1 \\ x_2 \\ x_3 \end{bmatrix} = \begin{bmatrix} 1 \\ 3 \\ -2 \end{bmatrix}$$

$$\begin{bmatrix} 1 & 2 & 3 \\ 0 & 1 & 5/2 \\ 0 & -5 & -7 \end{bmatrix} \begin{bmatrix} x_1 \\ x_2 \\ x_3 \end{bmatrix} = \begin{bmatrix} 1 \\ 3/2 \\ -2 \end{bmatrix}$$

$$\begin{bmatrix} 1 & 2 & 3 \\ 0 & 1 & 5/2 \\ 0 & 0 & 11/2 \end{bmatrix} \begin{bmatrix} x_1 \\ x_2 \\ x_3 \end{bmatrix} = \begin{bmatrix} 1 \\ 3/2 \\ 11/2 \end{bmatrix}$$

$$\begin{bmatrix} 1 & 2 & 3 \\ 0 & 1 & 5/2 \\ 0 & 0 & 1 \end{bmatrix} \begin{bmatrix} x_1 \\ x_2 \\ x_3 \end{bmatrix} = \begin{bmatrix} 1 \\ 3/2 \\ 1 \end{bmatrix}$$

$$\begin{bmatrix} 1 & 2 & 3 \\ 0 & 1 & 0 \\ 0 & 0 & 1 \end{bmatrix} \begin{bmatrix} x_1 \\ x_2 \\ x_3 \end{bmatrix} = \begin{bmatrix} 1 \\ -1 \\ 1 \end{bmatrix}$$

$$\begin{bmatrix} 1 & 0 & 0 \\ 0 & 1 & 0 \\ 0 & 0 & 1 \end{bmatrix} \begin{bmatrix} x_1 \\ x_2 \\ x_3 \end{bmatrix} = \begin{bmatrix} 0 \\ -1 \\ 1 \end{bmatrix}$$

Therefore the solution is given by $x_1 = 0, x_2 = -1, x_3 = 1$.

It appears that the Gauss elimination method should be the simplest of all methods from a computational point of view. Sometimes numerical analysts when performing Gauss's algorithm are quite satisfied when the matrix of coefficients is reduced into an upper triangular matrix with the diagonal elements not necessarily made equal to 1. The well-known back substitution is obviously applicable here, where we move upwards, calculating first x_n, then x_{n-1} using the value of x_n that we obtained. To calculate x_{n-2} we make use of x_n and x_{n-1} and so forth. The computation time in this case is less than for the case when A is transformed into a unit matrix; for it turns out that the amount of labour needed for back substitution is almost trivial when compared to the triangularization scheme. What is also nice about Gauss elimination is that one may use it to evaluate the determinant of A if needed (see Problem 3.8.25). One other remark about Gauss elimination is that one should be constantly aware of choosing the right pivotal strategy; one often meets a diagonal pivot too small in value compared with the element's values beneath it. In this case, the right thing to do to prevent numerical instability is to interchange the row containing a higher-value pivot. Wilkinson (1965), p. 216, provided a good example showing how this strategy ameliorates the solution. He considered the example

$$\begin{bmatrix} 0.000\,003 & 0.213\,472 & 0.332\,147 \\ 0.215\,512 & 0.375\,623 & 0.476\,625 \\ 0.173\,257 & 0.663\,257 & 0.625\,675 \end{bmatrix} \begin{bmatrix} x_1 \\ x_2 \\ x_3 \end{bmatrix} = \begin{bmatrix} 0.235\,262 \\ 0.127\,653 \\ 0.285\,321 \end{bmatrix}$$

and he worked this with a 6-decimal digit machine. If 0.000 003 were to be the pivot, the reduced system of equations becomes

$$\begin{bmatrix} 0.000\,003 & 0.213\,472 & 0.332\,147 \\ 0 & -15\,334.9 & -23\,860.0 \\ 0 & 12\,327.8 & -19\,181.7 \end{bmatrix} \begin{bmatrix} x_1 \\ x_2 \\ x_3 \end{bmatrix} = \begin{bmatrix} 0.235\,262 \\ -16\,900.5 \\ -13\,586.6 \end{bmatrix}$$

For the second elimination step, if −15 334.9 is the chosen pivot, we obtain the triangular system:

$$\begin{bmatrix} 0.000\,003 & 0.213\,472 & 0.332\,147 \\ 0 & -15\,334.9 & -23\,860.0 \\ 0 & 0 & -0.500\,000 \end{bmatrix} \begin{bmatrix} x_1 \\ x_2 \\ x_3 \end{bmatrix} = \begin{bmatrix} 0.235\,262 \\ -16\,900.5 \\ -0.200\,00 \end{bmatrix}$$

Using back substitution, we obtain:

$$x_3 = 0.400\,000, \quad x_2 = 0.479\,723, \quad x_1 = -1.333\,33$$

To see how poor these answers are, Wilkinson compared them with those obtained when a better pivotal strategy is followed. Going back to the equations, he selected 0.215 512 instead of 0.000 003, to give

$$\begin{bmatrix} 0.215\,512 & 0.375\,623 & 0.476\,625 \\ 0.000\,003 & 0.213\,472 & 0.332\,147 \\ 0.173\,257 & 0.663\,257 & 0.625\,675 \end{bmatrix} \begin{bmatrix} x_1 \\ x_2 \\ x_3 \end{bmatrix} = \begin{bmatrix} 0.127\,653 \\ 0.235\,262 \\ 0.285\,321 \end{bmatrix}$$

Executing the first elimination step, we obtain

$$\begin{bmatrix} 0.215\,512 & 0.375\,623 & 0.476\,625 \\ 0 & 0.213\,467 & 0.332\,140 \\ 0 & 0.361\,282 & 0.242\,501 \end{bmatrix} \begin{bmatrix} x_1 \\ x_2 \\ x_3 \end{bmatrix} = \begin{bmatrix} 0.127\,653 \\ 0.235\,260 \\ 0.182\,697 \end{bmatrix}$$

Again interchanging the second and third row to improve the pivotal strategy, we obtain, after the second elimination step,

$$\begin{bmatrix} 0.215\,512 & 0.375\,623 & 0.476\,625 \\ 0 & 0.361\,282 & 0.242\,501 \\ 0 & 0 & 0.188\,856 \end{bmatrix} \begin{bmatrix} x_1 \\ x_2 \\ x_3 \end{bmatrix} = \begin{bmatrix} 0.127\,653 \\ 0.182\,697 \\ 0.127\,312 \end{bmatrix}$$

Using back substitution, we find the solution to be

$$x_3 = 0.674\,122, \quad x_2 = 0.053\,205\,0, \quad x_1 = -0.991\,291$$

and since the correct answer, to ten figures, is

$$x_3 = 0.674\,121\,469\,4, \quad x_2 = 0.053\,203\,933\,91, \quad x_1 = -0.991\,289\,425\,2$$

we realize how a good pivotal strategy brings remarkable improvement in the accuracy.

The reader notices that the Gauss elimination scheme for triangularization with partial pivoting is a mathematical process equivalent to finding a permutation matrix P, an upper triangular matrix R and a lower triangular matrix L (with unities in the main diagonal) such that

$$PA = LR$$

and then solving the equations

$$Lz = Pb, \quad Rx = z$$

for x. Choosing P determines a strategy for interchanging the rows of A; finding L and R is part of the elimination process. Note that if the leading principal minors of A are non-zero and leading to good pivots during the triangularization scheme, P can be a unit matrix. Such a situation occurs when A is real symmetric or Hermitian and its leading principal minors are positive; A is called positive definite (see Section 4.6). Gauss elimination for positive definite matrices is a stable process without any pivotal strategy (see Wilkinson (1965)).

However in this case we can exploit the symmetry of A in order to economize on computer storage; it is here where Cholesky decomposition is introduced. To solve the real equations

$$Ax = b,$$

we compute an upper triangular matrix R, such that

$$A = R^T R$$

and solve

$$R^T y = b, \quad Rx = y$$

for x; see Exercise 3.10.3. Sometimes, it may be advisable to decompose A into the form

$$A = R^T DR,$$

where D is diagonal. This choice has, in some applications, better stability properties; see Peters and Wilkinson (1970).

Many other properties which exploit the symmetry of A have been devised; Bunch and Parlett (1971) survey these methods.

One other strategy that we should follow before starting the elimination algorithm is to scale A so that A becomes equilibrated. By an equilibrated matrix we mean one for which every row and every column has a length of order unity. We can almost invariably achieve this on any matrix; except that unfortunately the equilibrated form of a matrix is not unique; see Forsythe and Moler (1967), p. 45, allowing for different pivotal strategy when using Gauss elimination. Still equilibration is advantageous for it brings down the condition number of the matrix of coefficients, minimizing round-off errors, which are extremely important in ill-conditioned systems. As an example showing how scaling can be done, consider the matrix

$$A = \begin{bmatrix} 2 & 1 & 3 \times 10^6 \\ 1 & -1 & 10^6 \\ 1 & -2 & 0 \end{bmatrix}$$

If the columns are scaled we obtain the new matrix

$$A' = \begin{bmatrix} 0.2 & 0.1 & 0.3 \\ 0.1 & -0.1 & 0.1 \\ 0.1 & -0.2 & 0 \end{bmatrix}$$

The practical effects of various pivoting and equilibration strategies are discussed in Wilkinson (1965), Van der Sluis (1970) and Curtis and Reid (1972).

Whereas direct methods for solving linear equations yield solutions after a specified amount of computation, iterative methods, in contrast, start from an approximation of the solution and obtain after many repeated iterative cycles a better solution depending upon a prescribed accuracy. As we said before, iterative methods have the advantage of simplicity, uniformity and accuracy. They are either applied to problems for which convergence is known to be rapid or to large scale systems which the direct methods cannot deal with, especially to systems with sparse matrices for which for instance the elimination methods would need much storage. Such systems appear in vibrational problems or in partial differential equations. Besides, the method of Gauss is completely inaccurate since it involves much division apart from other arithmetical operations, and for large systems round-off errors become very frequent. Instead, iterative methods rely mainly on matrix multiplication. The only limitation on iterative methods is that the matrix of coefficients should possess properties to guarantee convergence; otherwise very scarce convergence or even divergence may be met. Of the known iterative methods, we mention the methods of Jacobi, and of Gauss–Siedel, and the relaxation methods, where the first method is largely of theoretical interest. In this method, the equations

$$Ax = b$$

are written in the following manner:

$$x_1 = \frac{b_1}{a_{11}} - \frac{a_{12}}{a_{11}} x_2 - \frac{a_{13}}{a_{11}} x_3 - \cdots - \frac{a_{1n}}{a_{11}} x_n$$

$$x_2 = \frac{b_2}{a_{22}} - \frac{a_{21}}{a_{22}} x_1 - \frac{a_{23}}{a_{22}} x_3 - \cdots - \frac{a_{2n}}{a_{22}} x_n$$

$$x_n = \frac{b_n}{a_{nn}} - \frac{a_{n1}}{a_{nn}} x_1 - \frac{a_{n2}}{a_{nn}} x_2 - \cdots - \frac{a_{n,n-1}}{a_{nn}} x_{n-1}$$

The method relies upon proposing an initial guess for the unknowns $x_1, \ldots, x_n$, and by substituting it in the right-hand side of the above equations, we obtain a better solution for the equations. Let the above equations be written in the form

$$x = \tilde{b} - \tilde{A}x$$

where

$$\tilde{b}^T = \left(\frac{b_1}{a_{11}}, \frac{b_2}{a_{22}}, \ldots, \frac{b_n}{a_{nn}} \right)$$

and

$$\tilde{A} = \begin{bmatrix} 0 & \frac{a_{12}}{a_{11}} & \frac{a_{13}}{a_{11}} & \cdots & \frac{a_{1n}}{a_{11}} \\ \frac{a_{21}}{a_{22}} & 0 & \frac{a_{23}}{a_{22}} & \cdots & \\ \vdots & & & & \\ \frac{a_{n1}}{a_{nn}} & & & & 0 \end{bmatrix}$$

Then by repetition of the iterative cycle several times, the solution is obtained within a prescribed accuracy. To prove the convergence of the method, let the iterations for x be successively written as follows:

$$x^1 = \tilde{b} - \tilde{A}x^0$$
$$x^2 = \tilde{b} - \tilde{A}x^1$$
$$\vdots$$
$$x^m = \tilde{b} - \tilde{A}x^{m-1}$$

And if x is the exact solution, i.e.

$$x = \tilde{b} - \tilde{A}x$$

we can easily see, upon subtraction, that

$$\begin{aligned} x^m - x &= \tilde{A}(x - x^{m-1}) \\ &= \tilde{A}^2(x^{m-2} - x) \\ &\vdots \\ &= (-)^m \tilde{A}^m (x^0 - x) \end{aligned}$$

Hence

$$\lim_{m\to\infty} (x^m - x) = 0$$

if

$$\lim_{m\to\infty} \tilde{A}^m = 0$$

In other words a necessary and sufficient condition for the convergence of the Jacobi method is that $\tilde{A}^m$ tends to zero as m tends to infinity. Such a limit occurs if the spectral radius of A is less than unity; see Exercise 4.1.33. For the moment, a sufficient condition can be that

$$\| \tilde{A} \| < 1$$

since

$$\| \tilde{A}^m \| \leqslant \| \tilde{A} \|^m$$

The reader can refer to Section 2.3 to see that a sufficient condition that $\| \tilde{A} \|$ is less than unity is that any one of the following conditions is satisfied:

1. $\displaystyle\sum_{\substack{j=1 \\ j\neq i}}^{n} \left| \frac{a_{ij}}{a_{ii}} \right| < 1, \quad \forall i = 1, \ldots, n$

2. $\displaystyle\sum_{\substack{i=1 \\ i\neq j}}^{n} \left| \frac{a_{ij}}{a_{ii}} \right| < 1, \quad \forall j = 1, \ldots, n$

3. $\displaystyle\sum_{\substack{i,j=1 \\ j\neq i}}^{n} \left| \frac{a_{ij}}{a_{ii}} \right|^2 < 1.$

The above conditions are based on $\| \tilde{A} \|_\infty$, $\| \tilde{A} \|_1$ and $\| \tilde{A} \|_E$ norms.

And an interesting substitution

$$x_i = p_i z_i$$

in the original system given by Faddeeva (1959), p. 121, where p_i are some positive numbers, can lead to another set of sufficient conditions; since both systems in x or z converge or diverge simultaneously. The reader will discover, by taking $p_i = 1/|a_{ii}|$, that any one of the following conditions can be sufficient for convergence of the original system:

1′. $$\sum_{\substack{j=1\\ j\neq i}}^{n} \left|\frac{a_{ij}}{a_{jj}}\right| < 1, \quad \forall i = 1, \ldots, n$$

2′. $$\sum_{\substack{i=1\\ i\neq j}}^{n} \left|\frac{a_{ij}}{a_{jj}}\right| < 1, \quad \forall j = 1, \ldots, n$$

3′. $$\sum_{\substack{i,j=1\\ i\neq j}}^{n} \left|\frac{a_{ij}}{a_{jj}}\right|^2 < 1,$$

Such conditions can be easily satisfied in many physical problems; for example in the solution of elliptic partial differential equations by the finite difference method one usually encounters a matrix of coefficients which is *diagonal dominant.* By a diagonal dominant matrix, we mean a matrix whose elements satisfy either

$$|a_{ii}| > \sum_{\substack{j=1\\ j\neq i}}^{n} |a_{ij}| \quad \text{or} \quad |a_{jj}| > \sum_{\substack{i=1\\ i\neq j}}^{n} |a_{ij}|$$

The above qualifications satisfy respectively conditions 1 and 2′. Such matrices when appearing in partial differential equations, in addition to being large and sparse, are nonsingular (see Exercise 4.1.22), assuring that there exists a unique solution for the vector x. For more properties of diagonal dominance, the reader is referred to Young and Gregory (1973), p. 1001. As an example of the Jacobi algorithm we repeat one examined by Young and Gregory (1973), p. 1010:

Example. Solve the equations

$$\begin{bmatrix} 4 & -1 & -1 & 0 \\ -1 & 4 & 0 & -1 \\ -1 & 0 & 4 & -1 \\ 0 & -1 & -1 & 4 \end{bmatrix} \begin{bmatrix} x_1 \\ x_2 \\ x_3 \\ x_4 \end{bmatrix} = \begin{bmatrix} 0 \\ 0 \\ 1000 \\ 1000 \end{bmatrix}$$

We write the equations in the form

$$\begin{aligned} x_1 &= \tfrac{1}{4}x_2 + \tfrac{1}{4}x_3 \\ x_2 &= \tfrac{1}{4}x_1 + \tfrac{1}{4}x_4 \\ x_3 &= 250 + \tfrac{1}{4}x_1 + \tfrac{1}{4}x_4 \\ x_4 &= 250 + \tfrac{1}{4}x_2 + \tfrac{1}{4}x_3 \end{aligned}$$

Choose a guess point $x_1 = x_2 = x_3 = x_4 = 0$ and proceed by iteration to obtain Table 1 (see p. 68), where n is the number of iterations. The solution is obtained in about 23 iterations.

The Gauss–Siedel method is similar to that of Jacobi, except that it is more practical. Once x_1 is calculated, its new value enters into the calculation of x_2 and so forth. The result is less storage and faster convergence. For convergence proof, the reader is referred to Exercise 4.1.29; where the diagonal dominance property of A is made use of.

Example. Consider the previous example of Young and Gregory (1973), p. 1017. Using the Gauss–Siedel method, we obtain Table 2 (see p. 68).
where n is the number of iterations. The solution is reached after about 13 iterations, showing the faster convergence compared to that of the Jacobi method.

The relaxation method is perhaps the most popular of the iterative methods. Let the equations be written in the form

$$Ax - b = \delta$$

where δ is a residual error vector and is equal to zero only if x is the exact solution. Let x^0 be an initial guess for the solution; then

$$Ax^0 - b = \delta^0$$

Choose

$$x^1 = x^0 + \Delta x^0,$$

such that when

$$Ax^1 - b = \delta^1$$

then

$$\| \delta^1 \| \leqslant \| \delta^0 \|$$

The problem becomes that of calculating Δx^0, such that the above inequality holds. From the above equations we obtain

$$\delta^1 = A(x^0 + \Delta x^0) - b = A\Delta x^0 + \delta^0$$

Therefore it is required to choose a value for Δx^0, such that

$$\| A\Delta x^0 + \delta^0 \| \leqslant \| \delta^0 \|$$

Let δ_k^0 be the largest element in magnitude in δ^0. Choose Δx^0 as follows

$$\Delta x^{0^T} = [0, 0, \ldots, 0, \Delta x_k^0, 0, \ldots, 0]$$

such that

$$\Delta x_k^0 = -\frac{\delta_k^0}{a_{kk}}$$

Table 1

n	x_1	x_2	x_3	x_4	Error in x_1	Error in x_2	Error in x_3	Error in x_4
0	0	0	0	0	−125	−125	−375	−375
1	0	0	250	250	−125	−125	−125	−125
2	62.5	62.5	312.5	312.5	−62.5	−62.5	−62.5	−62.5
3	93.7	93.7	343.75	343.75	−31.25	−31.25	−31.25	−31.25
4	10.934 5	109.375	359.375	359.375	−15.625	−15.625	−15.652	−15.625
⋮								
23	125	125	375	375	0	0	0	0

Table 2

n	x_1	x_2	x_3	x_4	Error in x_1	Error in x_2	Error in x_3	Error in x_4
0	0	0	0	0	−125	−125	−125	−125
1	0	0	250	312.5	−125	−125	−125	−62.5
2	62.5	93.75	343.375	359.375	−62.5	−31.25	−31.25	−15.625
3	109.375	117.187 5	367.187 5	371.093 7	−15.625	−7.812 5	−7.812 5	−3.906 3
4	121.093 7	123.046 9	373.046 9	374.023 4	−3.906 3	−1.953 1	−1.953 1	−0.976 6
⋮								
13	125	125	375	375	0	0	0	0

By such a choice, the iterative process will converge, since

$$\| A\Delta x^0 + \delta^0 \|_1 = \sum_{i \neq k} \left| \delta_i^0 - \frac{a_{ik}\,\delta_k^0}{a_{kk}} \right|$$

$$= \sum_{i \neq k} \left| \delta_i^0 - \frac{a_{ik}\,\delta_k^0}{a_{kk}} \right| + |\delta_k^0| - |\delta_k^0|$$

$$\leqslant \sum_{i \neq k} |\delta_i^0| + |\delta_k^0| \sum_{i \neq k} \left| \frac{a_{ik}}{a_{kk}} \right| + |\delta_k^0| - |\delta_k^0|$$

$$= \| \delta^0 \|_1 - |\delta_k^0| \left(1 - \sum_{i \neq k} \left| \frac{a_{ik}}{a_{kk}} \right| \right)$$

If A is diagonal dominant, the bracket on the right-hand side of the above inequality is positive, and we finally obtain

$$\| A\Delta x^0 + \delta^0 \|_1 \leqslant \| \delta^0 \|_1$$

which asserts the convergence of the relaxation method.

The method of relaxation in its simplest form relies upon reducing the numerically largest residual to zero at each step. It will terminate when all residuals of the last equation are equal to zero. Although the method is not very practical on machine computation, it is very easy in hand computation. For the number of operations at each step is small, in comparison with the previous methods. It is only natural, therefore, that it will require more iterations to reach the solution, than for example the Gauss–Siedel method. For this reason, it should not be criticized since a desk calculator can be very effective with the relaxation method and very troublesome with the Gauss–Siedel method.

Example. We treat the same problem

$$\begin{bmatrix} -4 & -1 & -1 & 0 \\ -1 & 4 & 0 & -1 \\ -1 & 0 & 4 & -1 \\ 0 & -1 & -1 & 4 \end{bmatrix} \begin{bmatrix} x_1 \\ x_2 \\ x_3 \\ x_4 \end{bmatrix} - \begin{bmatrix} 0 \\ 0 \\ 1000 \\ 1000 \end{bmatrix} = \delta$$

Choose

$$x_1 = x_2 = x_3 = x_4 = 0$$

giving

$$\delta^{0T} = [0, 0, -1000, -1000]$$

Next choose

$$x_1 = 0, \quad x_2 = 0, \quad x_3 = 0 + \frac{1000}{4} = 250, \quad x_4 = 0$$

giving

$$\delta^{1^T} = [-250, 0, 0, -1250]$$

Next choose

$$x_1 = 0, \quad x_2 = 0, \quad x_3 = 250, \quad x_4 = 0 + \frac{1250}{4} = 312.5$$

giving

$$\delta^{2^T} = [-250, -312.5, -312.5, 0]$$

Next choose

$$x_1 = 0, \quad x_2 = 0, \quad x_3 = 250 + \frac{312.5}{4} = 328.1, \quad x_4 = 312.5$$

giving

$$\delta^{3^T} = [-328.1, -312.5, 0, -78.1]$$

Next choose

$$x_1 = 0 + \frac{328.1}{4} = 82, \quad x_2 = 0, \quad x_3 = 328.1, \quad x_4 = 312.5$$

giving

$$\delta^{4^T} = [0, -394.5, -82, -78.1]$$

Next choose

$$x_1 = 82, \quad x_2 = 0 + \frac{394.5}{4} = 98.6, \quad x_3 = 328.1, \quad x_4 = 312.5$$

giving

$$\delta^{5^T} = [-98.6, 0, -82, -176.7]$$

The reader can follow the same steps, and make sure that the norm of the vector δ decreases, meaning that we are approaching the solution.

There are many other iterative methods for solving linear equations. Young and Gregory (1973), pp. 1026–1074, treated mainly the successive over-relaxation method and the Peaceman–Rachford alternating direction implicit method. Stoer and Bulirsch (1980), Chapter 8, considered some others, and made a comparison between available methods.

(b) *A is singular or rectangular with $m \leqslant n$*

Let the system of equations be written in the following form:

$$\begin{bmatrix} a_{11} & a_{12} & \cdots & a_{1n} \\ a_{21} & a_{22} & & \\ \vdots & & & \\ a_{m1} & & & a_{m,n} \end{bmatrix} \begin{bmatrix} x_1 \\ x_2 \\ \vdots \\ x_n \end{bmatrix} = \begin{bmatrix} b_1 \\ b_2 \\ \vdots \\ b_m \end{bmatrix}$$

We operate on the above equations by a series of elementary row operations until the matrix A of coefficients is transformed into an echelon form like the following

$$\begin{bmatrix} \text{Echelon form} \\ \hdashline O_{m-\rho,n} \end{bmatrix} \begin{bmatrix} x_1 \\ x_2 \\ \vdots \\ x_n \end{bmatrix} = \begin{bmatrix} b'_1 \\ \vdots \\ b'_\rho \\ \hdashline b'_{\rho+1} \\ \vdots \\ b'_m \end{bmatrix}$$

The vector b will be consequently altered to b' as shown. It becomes obvious that for the above equations to have a solution, we must have:

$$b'_{\rho+1} = b'_{\rho+2} = \cdots = b'_m = 0$$

This can be stated differently by the following theorem:

THEOREM. The set of linear simultaneous equations $Ax = b$ has a solution for x if and only if

$$\rho(A) = \rho(A \mid b)$$

The proof of the above theorem is left as an exercise for the reader.

Next, by operating on the matrix A by a series of elementary column operations of the type P_1 only, the equations can be reduced to the following form:

$$\begin{bmatrix} I_{\rho,\rho} & \vdots & Q_{\rho,n-\rho} \\ \hdashline & O_{m-\rho,n} & \end{bmatrix} \begin{bmatrix} x'_1 \\ \vdots \\ x'_\rho \\ \hdashline x'_{\rho+1} \\ \vdots \\ x'_n \end{bmatrix} = \begin{bmatrix} b'_1 \\ \vdots \\ b'_\rho \\ \hdashline 0 \end{bmatrix}$$

Now the solution of the above equations can be directly obtained as follows:

$$\begin{bmatrix} x'_1 \\ \vdots \\ x'_\rho \end{bmatrix} = \begin{bmatrix} b'_1 \\ \vdots \\ b'_\rho \end{bmatrix} - Q_{\rho,n-\rho} \begin{bmatrix} x'_{\rho+1} \\ \vdots \\ x'_n \end{bmatrix}$$

and $x'_{\rho+1}, \ldots, x'_n$ can assume any arbitrary values. The solution can also be put in a much more convenient form by taking

$$\begin{bmatrix} x'_{\rho+1} \\ \vdots \\ x'_n \end{bmatrix} = - \begin{bmatrix} c_1 \\ \vdots \\ c_{n-\rho} \end{bmatrix} \quad \text{and } Q_{\rho,n-\rho} = [q^1 \mid q^2 \mid \cdots \mid q^{n-\rho}]$$

to read

$$\begin{bmatrix} x'_1 \\ x'_2 \\ \vdots \\ x'_\rho \\ \hdashline x'_{\rho+1} \\ \vdots \\ x'_n \end{bmatrix} = \begin{bmatrix} b'_1 \\ \vdots \\ b'_\rho \\ \hdashline 0 \\ 0 \\ \vdots \\ 0 \end{bmatrix} + c_1 \begin{bmatrix} \\ q^1 \\ \\ \hdashline -1 \\ 0 \\ \\ 0 \end{bmatrix} + c_2 \begin{bmatrix} \\ q^2 \\ \\ \hdashline 0 \\ -1 \\ \\ 0 \end{bmatrix} + \cdots + c_{n-\rho} \begin{bmatrix} \\ q^{n-\rho} \\ \\ \hdashline 0 \\ \vdots \\ -1 \end{bmatrix}$$

where the scalars $c_1, \ldots, c_{n-\rho}$ can take any arbitrary values.

If on the other hand $b = 0$, the same approach is valid and there always exists a solution since

$$\rho(A) = \rho(A \mid 0)$$

In this case the solution is the same as above except that

$$b'_1 = b'_2 = \cdots = b'_\rho = 0.$$

To this end one must note that the number of linearly independent solutions for the homogeneous equations

$$Ax = 0$$

is equal to $n - \rho$; in other words it is equal to the number of unknowns minus the rank of A.

The solution of the non-homogeneous equations can also be written in the form

$$\begin{bmatrix} x'_1 \\ \vdots \\ x'_\rho \\ \hdashline x'_{\rho+1} \\ \vdots \\ x'_n \end{bmatrix} = \begin{bmatrix} b'_1 \\ \vdots \\ b'_\rho \\ \hdashline \\ 0 \\ \\ \end{bmatrix} + \begin{bmatrix} Q_{\rho,n-\rho} \\ \hdashline -I_{n-\rho,n-\rho} \end{bmatrix} \begin{bmatrix} c_1 \\ \vdots \\ c_{n-\rho} \end{bmatrix}$$

where A must be orthogonal to the matrix in the right-hand side so that we come back to $Ax = b$, i.e.

$$A \begin{bmatrix} Q_{\rho,n-\rho} \\ \hdashline -I_{n-\rho,n-\rho} \end{bmatrix} = O_{m,n-\rho}$$

The proof of the above result can be an exercise for the reader.

Example. Solve the equations

$$\begin{bmatrix} 10 & 0 & 5 & 0 \\ 4 & -2 & 2 & 1 \\ 2 & 4 & 1 & -2 \end{bmatrix} \begin{bmatrix} x_1 \\ x_2 \\ x_3 \\ x_4 \end{bmatrix} = \begin{bmatrix} 1 \\ 2 \\ -3 \end{bmatrix}$$

Operating on the equations by a series of elementary row operations we obtain

$$\left[\begin{array}{cc:cc} 1 & 0 & \tfrac{1}{2} & 0 \\ 0 & 1 & 0 & -\tfrac{1}{2} \\ \hdashline 0 & 0 & 0 & 0 \end{array}\right] \begin{bmatrix} x_1 \\ x_2 \\ x_3 \\ x_4 \end{bmatrix} = \begin{bmatrix} 1/10 \\ -4/5 \\ \hdashline 0 \end{bmatrix}$$

The equations have a solution, since they are found consistent, and it is given by

$$\begin{bmatrix} x_1 \\ x_2 \\ x_3 \\ x_4 \end{bmatrix} = \begin{bmatrix} 1/10 \\ -4/5 \\ 0 \\ 0 \end{bmatrix} + c_1 \begin{bmatrix} 1/2 \\ 0 \\ -1 \\ 0 \end{bmatrix} + c_2 \begin{bmatrix} 0 \\ -\tfrac{1}{2} \\ 0 \\ -1 \end{bmatrix}$$

where c_1 and c_2 can take any arbitrary values. The reader can check that

$$A \begin{bmatrix} 1/10 \\ -4/5 \\ 0 \\ 0 \end{bmatrix} = \begin{bmatrix} 1 \\ 2 \\ -3 \end{bmatrix} \quad \text{and} \quad A \begin{bmatrix} \tfrac{1}{2} & 0 \\ 0 & -\tfrac{1}{2} \\ -1 & 0 \\ 0 & -1 \end{bmatrix} = 0$$

To this end, one must note that we do not know whether a square matrix is singular or nonsingular before we run successively our chosen algorithm. This explains why the Gauss elimination method has gained great popularity – it can be applied to both types of matrix.

(c) *A is rectangular with $m > n$*
For the equations

$$Ax = b$$

with A of order (m, n) and with $m > n$, there is no exact solution because

$$\rho(A) \neq \rho(A \,\vdots\, b)$$

in general. If it happens as a special case that

$$\rho(A) = \rho(A \,\vdots\, b)$$

then the problem is treated as in the previous cases. When both sides of the above relation are unequal, as is usually so, the equations have only an approximate solution, which minimizes the error between Ax and b; i.e. the quantity

$$\| Ax - b \|_p$$

The easiest case is when $p = 2$, and the problem is then to solve for x:

$$\min_x \| Ax - b \|_2 \quad \text{or} \quad \min_x \| Ax - b \|_2^2$$

This is equivalent, if A and b are real, to solving:

$$\min_x (Ax - b)^T(Ax - b)$$

The solution x is found from the equation

$$\frac{\partial}{\partial x}(Ax - b)^T(Ax - b) = \frac{\partial}{\partial x}(x^T A^T Ax + b^T b - 2x^T A^T b) = 0;$$

giving

$$A^T Ax = A^T b$$

The matrix on the left-hand side is nonsingular if $\rho(A) = n$, as the reader can show. If $\rho(A) < n$, the matrix $A^T A$ is singular, and the solution of the least-squares problem becomes complicated. Peters and Wilkinson (1970) and Hanson and Lawson (1969) discuss this problem in detail; see also Section 5.3.

Example. Solve the equations

$$\begin{bmatrix} 1 & 2 \\ 0 & 3 \\ -1 & 1 \\ 2 & -1 \end{bmatrix} \begin{bmatrix} x_1 \\ x_2 \end{bmatrix} = \begin{bmatrix} 1 \\ 2 \\ -1 \\ 3 \end{bmatrix}$$

$$A^T A = \begin{bmatrix} 1 & 0 & -1 & 2 \\ 2 & 3 & 1 & -1 \end{bmatrix} \begin{bmatrix} 1 & 2 \\ 0 & 3 \\ -1 & 1 \\ 2 & -1 \end{bmatrix} = \begin{bmatrix} 6 & -1 \\ -1 & 15 \end{bmatrix}$$

$$A^T b = \begin{bmatrix} 1 & 0 & -1 & 2 \\ 2 & 3 & 1 & -1 \end{bmatrix} \begin{bmatrix} 1 \\ 2 \\ -1 \\ 3 \end{bmatrix} = \begin{bmatrix} 8 \\ 4 \end{bmatrix}$$

Hence

$$\begin{bmatrix} 6 & -1 \\ -1 & 15 \end{bmatrix} \begin{bmatrix} x_1 \\ x_2 \end{bmatrix} = \begin{bmatrix} 8 \\ 4 \end{bmatrix}$$

giving

$$x_1 = \frac{124}{89}, \quad x_2 = \frac{32}{89}$$

3.10. Application to nonlinear equations

The set of nonlinear simultaneous equations

$$x_i = f_i(x_1, x_2, \ldots, x_n), \quad i = 1, 2, \ldots, n$$

can be treated by an iterative method similar to the Jacobi method. Let x^0 be an initial guessing solution of the set of equations, then

$$x^1 = f(x^0)$$

where

$$f^T = [f_1 f_2 \ldots f_n]$$

and

$$x^{0^T} = [x_1^0, x_2^0, \ldots, x_n^0]$$

x^1 represents a first approximation to the solution. If the iteration process is repeated successively, we obtain a convergence of the method if

$$\left\| \frac{\partial f}{\partial x} \right\| < 1$$

To prove this, we write the iterations as follows:

$$\begin{aligned} x^1 &= f(x^0) \\ x^2 &= f(x^1) \\ &\vdots \\ x^m &= f(x^{m-1}) \end{aligned}$$

A solution is reached if

$$x = f(x)$$

Subtracting the equations we obtain

$$x^m - x = f(x^{m-1}) - f(x)$$

$$= \frac{\partial f}{\partial x}\bigg| (x^{m-1} - x) = \frac{\partial f}{\partial x}\bigg| \cdot \frac{\partial f}{\partial x}\bigg| (x^{m-2} - x)$$

$$x \leqslant \epsilon^{m-1} \leqslant x^{m-1} \qquad x \leqslant \epsilon^{m-1} \leqslant x^{m-1} \qquad x \leqslant \epsilon^{m-2} \leqslant x^{m-2}$$

$$\vdots$$

$$\approx \left[\frac{\partial f}{\partial x}\bigg| \right]^m (x^0 - x)$$

$$x \leqslant \epsilon \leqslant x^{m-1}$$

$$x \leqslant \epsilon \leqslant x^{m-2}$$

$$x \leqslant \epsilon \leqslant x^0$$

Hence

$$\lim_{m \to \infty} (x^m - x) = 0$$

if

$$\left\| \frac{\partial f}{\partial x}\bigg|_\epsilon \right\| < 1$$

Example. Solve the equations (Carnahan, Luther and Wilkes (1969), p. 308)

$$½ \sin(xy) - \frac{y}{4\pi} - \frac{x}{2} = 0, \quad \left(1 - \frac{1}{4\pi}\right)(e^{2x} - e) + \frac{ey}{\pi} - 2ex = 0$$

To solve the above equations, we rewrite them as follows:

$$x = \sin(xy) - \frac{y}{2\pi}$$

$$y = 2\pi x - (\pi - ¼)(e^{2x-1} - 1)$$

Choose as a first guess:

$$x = 0.4, \quad y = 3$$

Within slide rule accuracy, the Jacobi-type iteration gives

$$x = \sin(1.2) - \frac{3}{2\pi} = 0.455$$

$$y = 2\pi(0.4) - 2.89(e^{-0.2} - 1) = 3.03$$

By another iteration we obtain

$$x = \sin(1.379) - \frac{3.03}{2\pi} = 0.499$$

$$y = 2\pi(0.455) - 2.89(e^{-0.09} - 1) = 3.11$$

A third iteration gives

$$x = \sin(1.552) - \frac{3.11}{2\pi} = 0.505$$

$$y = 2\pi(0.499) - 2.89(e^{-0.002} - 1) = 3.14$$

And a fourth iteration gives

$$x = \sin(1.585) - \frac{3.14}{2\pi} = 0.5$$

$$y = 2\pi(0.505) - 2.89(e^{0.01} - 1) = 3.14$$

By a fifth iteration we obtain

$$x = \sin(1.57) - \frac{3.14}{2\pi} = 0.5$$

$$y = 2\pi(0.5) - 2.89(e^{0} - 1) = \pi$$

which shows that the iteration process converges.

It is to be noted that the set of nonlinear equations could well be treated using the Gauss–Siedel method to accelerate the convergence.

Exercises 3.10

1. Show that, if A is nonsingular, $Ax = b$ has a unique solution for x. *Hint*: write $Ax = b, Ay = b$; then subtract.
2. Find the condition for which the system of linear simultaneous equations

$$\begin{bmatrix} A & B \\ C & D \end{bmatrix}\begin{bmatrix} x \\ u \end{bmatrix} = \begin{bmatrix} v \\ y \end{bmatrix}$$

has a unique solution for x and y. A, B, C and D are given matrices with A and D being square, and u and v are given vectors. Hence solve the equations

$$\begin{bmatrix} 1 & 1 & 0 & 1 & 8 \\ -1 & 1 & 2 & -1 & 0 \\ -2 & 0 & 4 & 6 & 2 \\ 0 & -3 & -1 & 1 & 4 \\ 3 & 1 & 2 & 5 & -1 \end{bmatrix}\begin{bmatrix} x_1 \\ x_2 \\ x_3 \\ 0 \\ 1 \end{bmatrix} = \begin{bmatrix} 2 \\ -1 \\ 2 \\ x_4 \\ x_5 \end{bmatrix}$$

3. Decompose the matrix

$$A = \begin{bmatrix} 5 & -2 & -4 \\ -2 & 2 & 2 \\ -4 & 2 & 5 \end{bmatrix}$$

into the form $A = R^T R$, where R is an upper triangular matrix. *Hint*: $r_{11} = \sqrt{a_{11}}$,

$$r_{1j} = a_{1j}/r_{11}\ (j > 1),\ r_{ii} = \sqrt{a_{ii} - \sum_{k=1}^{i-1} r_{ki}^2}\ (1 < i \leqslant n),\ r_{ij} = \left(a_{ij} - \sum_{k=1}^{i-1} r_{ki} r_{kj}\right) / r_{ii}\ (i < j),$$

$r_{ij} = 0\ (i > j)$. Hence solve $Ax = b$, with $b^T = [1, 2, 1]$, by first solving $R^T y = b$ then solving $Rx = y$. Write an algorithm for obtaining successively y_i and x_i. For more information about the Cholesky method, refer to Stoer and Bulirsch (1980), p. 172.

4. Show how the Gauss–Jordan method explained in Exercise 3.8.16 can be used to solve a set of linear simultaneous equations $Ax = b$. Apply the method to solve the following equations:

$$x_1 + x_2 + x_3 + x_4 = 1, \quad x_1 + x_2 + x_3 - x_4 = 2,$$
$$x_1 + x_2 - x_3 - x_4 = 3, \quad x_1 - x_2 - x_3 - x_4 = 4$$

5. Solve, using one of the iterative methods,

$$6x_1 + 2x_2 + 3x_3 = 10$$
$$2x_1 + 6x_2 + x_3 - x_4 = 20$$
$$-x_1 + 2x_2 + 5x_3 - x_4 = 10$$
$$2x_1 + 2x_2 + 3x_3 + 10x_4 = 20$$

6. The finite difference method can be used to solve a partial differential equation in a region R by dividing the region into subregions using a rectangular grid if the dependent variable u is given in terms of $u(x, y)$ or a circular grid if u is given in terms of $u(r, \theta)$. Then, by representing the operators $\partial u/\partial x$, $\partial u/\partial y$, $\Delta^2 u$ in terms of the function u at different points around $u(x, y)$, and by substituting into the partial differential equation we obtain a set of linear equations to solve for u inside R. For example, to solve for u Poisson's equation

$$\frac{\partial^2 u}{\partial x^2} + \frac{\partial^2 u}{\partial y^2} + \frac{P}{s} = 0$$

inside a square region, we substitute

$$\frac{\partial^2 u}{\partial x^2} + \frac{\partial^2 u}{\partial y^2} = \frac{u_r + u_l + u_u + u_b - 4u_0}{h^2}$$

where u_r, u_l, u_u, u_b are respectively the values of u to the right, left, upper and bottom of u_0 and are spaced by a distance h from u_0, in order to obtain a set of linear equations to solve for u inside R. The variable u may represent the deflection of a square stretched membrane of side length l fixed at its four boundaries. P is the pressure on the membrane and s is the tension per unit length of the membrane. Show that the set of linear equations obtained have a diagonal dominant matrix of coefficients. Hence obtain u by using a suitable numerical method if the grid divides the region R into 16 smaller squares. Make use of the symmetry of u.

7. Find the condition for the system of equations

$$\alpha_{nb} i = I, \quad \beta_{lb} v = E, \quad v = Ri, \quad (n + l = b)$$

to have a unique solution for i and v. Hence or otherwise solve the equations

$$\begin{bmatrix} 1 & 1 & 0 & 0 & 1 \\ 0 & 1 & -1 & -1 & 1 \end{bmatrix} \begin{bmatrix} i_1 \\ i_2 \\ i_3 \\ i_4 \\ i_5 \end{bmatrix} = \begin{bmatrix} e^t \\ 0 \end{bmatrix},$$

$$\begin{bmatrix} 0 & 1 & 0 & 0 & -1 \\ -1 & 1 & 1 & 0 & 0 \\ 0 & 0 & 1 & -1 & 0 \end{bmatrix} \begin{bmatrix} v_1 \\ v_2 \\ v_3 \\ v_4 \\ v_5 \end{bmatrix} = \begin{bmatrix} e^t \\ 0 \\ 0 \end{bmatrix}$$

if $v_j = 2i_j$, for $j = 1, 2, 3, 4, 5$.

8. Show that, for the system of linear equations $a_1x_1 + a_2x_2 + a_3x_3 = 0$ and $b_1x_1 + b_2x_2 + b_3x_3 = 0$,

$$\frac{x_1}{\begin{vmatrix} a_2 & a_3 \\ b_2 & b_3 \end{vmatrix}} = \frac{-x_2}{\begin{vmatrix} a_1 & a_3 \\ b_1 & b_3 \end{vmatrix}} = \frac{x_3}{\begin{vmatrix} a_1 & a_2 \\ b_1 & b_2 \end{vmatrix}}$$

9. Find the condition for the points (x_1, y_1), (x_2, y_2), (x_3, y_3) and (x_4, y_4) to lie on a straight line.

10. Find the condition for the system of linear simultaneous equations

$$\begin{bmatrix} A & B \\ C & D \end{bmatrix} \begin{bmatrix} x \\ u \end{bmatrix} = \begin{bmatrix} v \\ y \end{bmatrix}$$

to have a solution for x and y. A, B, C and D are given matrices with A and D being square and u and v are given vectors. Hence find the value of α for which the following system of equations has a solution and find the solution

$$\begin{bmatrix} 1 & 0 & 2 & 1 & -1 \\ 0 & 1 & 1 & 0 & 1 \\ -2 & 1 & -3 & 1 & 1 \\ 1 & 0 & 0 & -1 & 2 \\ 0 & 1 & 1 & 1 & -1 \end{bmatrix} \begin{bmatrix} x_1 \\ x_2 \\ x_3 \\ 1 \\ 2 \end{bmatrix} = \begin{bmatrix} 1 \\ 0 \\ \alpha \\ x_4 \\ x_5 \end{bmatrix}$$

11. Find the condition for the system of equations

$$\begin{bmatrix} A & 0 \\ B & C \end{bmatrix} \begin{bmatrix} x \\ y \end{bmatrix} = \begin{bmatrix} u \\ v \end{bmatrix}$$

to have a unique solution for x and more than one solution for y.

12. Find the value of α for the system of linear equations

$$\begin{bmatrix} 5 & -2 & -4 & 0 & 0 \\ -2 & 2 & 2 & 0 & 0 \\ -4 & 2 & 5 & 0 & 0 \\ 1 & -1 & 0 & 1 & 2 \\ -3 & 2 & 1 & 0 & 1 \end{bmatrix} \begin{bmatrix} x \\ y \\ z \\ u \\ v \end{bmatrix} = \alpha \begin{bmatrix} x \\ y \\ z \\ 1 \\ -1 \end{bmatrix}$$

to have a solution for x, y, z, u and v, hence find the solution.

13. Find the value of λ for which the system of equations

$$\lambda x + y + z = 4 - \lambda, \quad x + \lambda y + z = 2 + \lambda, \quad x + y + \lambda z = 3$$

has

1. a unique solution for x, y and z.
2. more than one solution for x, y and z.
3. no solution for x, y and z.

14. Show that the equation of the straight line fitting the points

x_1	x_2	...	x_n
y_1	y_2	...	y_n

is $y = ax + b$, where a and b are given from

$$\begin{bmatrix} \sum_{i=1}^{n} x_i^2 & \sum_{i=1}^{n} x_i \\ \sum_{i=1}^{n} x_i & n \end{bmatrix} \begin{bmatrix} a \\ b \end{bmatrix} = \begin{bmatrix} \sum_{i=1}^{n} x_i y_i \\ \sum_{i=1}^{n} y_i \end{bmatrix}$$

CHAPTER FOUR

THE MATRIX EIGENVALUE PROBLEM

4.1. Characteristic roots and vectors

In many physical applications one often encounters the following problem: given a square matrix A of order n, determine the value of a scalar λ and a non-zero vector u which satisfy the equation

$$Au = \lambda u$$

This problem is called the *eigenvalue problem* for the matrix A. λ is called the *eigenvalue* (proper value, characteristic value) of A and u its *eigenvector* (proper vector, characteristic vector). The above equation can also be written in the form

$$(\lambda I - A)\, u = 0$$

This system of n linear simultaneous equations in u has a non-trivial solution for the vector u only if the matrix $(\lambda I - A)$ is singular, i.e.

$$| \lambda I - A | = 0$$

The expansion of this determinant $| \lambda I - A |$ yields a polynomial in λ of degree equal to n, which is called the *characteristic polynomial* of the matrix A. The n roots of the above equation, called the *characteristic equation*, are the eigenvalues of A. The set of all eigenvalues is called the *spectrum* of A.

Therefore the eigenvalue problem has a non-trivial solution for u only if λ takes the value of one of the roots of the characteristic equation.

To each eigenvalue λ_i there corresponds a solution for the vector u. Therefore there are n eigenvectors of A.

Example. Solve the eigenvalue problem

$$\begin{bmatrix} 1 & 0 & -2 \\ 1 & 1 & 0 \\ 0 & 1 & 2 \end{bmatrix} u = \lambda u$$

Grouping terms of u we obtain

$$\begin{bmatrix} \lambda - 1 & 0 & 2 \\ -1 & \lambda - 1 & 0 \\ 0 & -1 & \lambda - 2 \end{bmatrix} u = 0$$

Hence $u \neq 0$ only if the determinant of the matrix of the left-hand side is zero, i.e. only if

$$(\lambda - 1)(\lambda - 1)(\lambda - 2) + 2 = 0,$$

giving

$$\lambda(\lambda^2 - 4\lambda + 5) = 0$$

Hence the eigenvalues are

$$\lambda_1 = 0, \quad \lambda_2 = 2 + i, \quad \lambda_3 = 2 - i, \quad i^2 = -1$$

To obtain the eigenvectors of A we solve the set of linear simultaneous equations

$$(\lambda_i I - A)\, u^i = 0$$

after putting $\lambda_i = \lambda_1, \lambda_2, \lambda_3$. Hence to calculate u^1, we solve

$$(\lambda_1 I - A)\, u^1 = 0$$

where $\lambda_1 = 0$.

Thus

$$\begin{bmatrix} -1 & 0 & 2 \\ -1 & -1 & 0 \\ 0 & -1 & -2 \end{bmatrix} u^1 = 0 \Rightarrow u^1 = \begin{bmatrix} -2 \\ 2 \\ -1 \end{bmatrix}$$

To calculate u^2, we solve

$$(\lambda_2 I - A)\, u^2 = 0$$

with $\lambda_2 = 2 + i$ and obtain

$$\begin{bmatrix} 1+i & 0 & 2 \\ -1 & 1+i & 0 \\ 0 & -1 & i \end{bmatrix} u^2 = 0 \Rightarrow u^2 = \begin{bmatrix} 1-i \\ -i \\ -1 \end{bmatrix}$$

Finally, for u^3, we solve

$$(\lambda_3 I - A)\, u^3 = 0$$

with $\lambda_3 = 2 - i$ and obtain

$$\begin{bmatrix} 1-i & 0 & 2 \\ -1 & 1-i & 0 \\ 0 & -1 & -i \end{bmatrix} u^3 = 0 \Rightarrow u^3 = \begin{bmatrix} 1+i \\ i \\ -1 \end{bmatrix}$$

Example. Solve the eigenvalue problem

$$\begin{bmatrix} 5 & -2 & -4 \\ -2 & 2 & 2 \\ -4 & 2 & 5 \end{bmatrix} u = \lambda u$$

From the equations

$$\begin{bmatrix} \lambda - 5 & 2 & 4 \\ 2 & \lambda - 2 & -2 \\ 4 & -2 & \lambda - 5 \end{bmatrix} u = 0$$

we must have, for $u \neq 0$, the determinant of the matrix of the left-hand side equal to zero, giving

$$\lambda^3 - 12\lambda^2 + 21\lambda - 10 = 0$$

Solving, we get $\lambda_1 = 10, \lambda_2 = 1, \lambda_3 = 1$.
u^1 is obtained by solving

$$\begin{bmatrix} 5 & 2 & 4 \\ 2 & 8 & -2 \\ 4 & -2 & 5 \end{bmatrix} u^1 = 0 \Rightarrow u^1 = \begin{bmatrix} 1 \\ -\tfrac{1}{2} \\ -1 \end{bmatrix}$$

u^2 and u^3 are obtained by solving

$$\begin{bmatrix} -4 & 2 & 4 \\ 2 & -1 & -2 \\ 4 & -2 & -4 \end{bmatrix} u = 0 \Rightarrow u^2 = \begin{bmatrix} -\tfrac{1}{2} \\ -1 \\ 0 \end{bmatrix}, \quad u^3 = \begin{bmatrix} -1 \\ 0 \\ -1 \end{bmatrix}$$

Example. Solve the eigenvalue problem

$$\begin{bmatrix} 1 & 1 \\ 0 & 1 \end{bmatrix} u = \lambda u$$

From the equations

$$\begin{bmatrix} \lambda - 1 & -1 \\ 0 & \lambda - 1 \end{bmatrix} u = 0$$

We must have, for $u \neq 0$

$$\begin{vmatrix} \lambda - 1 & -1 \\ 0 & \lambda - 1 \end{vmatrix} = 0 \Rightarrow \lambda_1 = \lambda_2 = 1$$

u^1 and u^2 are obtained by solving

$$\begin{bmatrix} 0 & -1 \\ 0 & 0 \end{bmatrix} u = 0 \Rightarrow u^1 = u^2 = \begin{bmatrix} -1 \\ 0 \end{bmatrix}$$

Each matrix A in the last three examples had a different nature. In the first example, the eigenvalues were distinct and so the eigenvectors were found to be linearly independent. This is the result of a theorem which we are going to prove.

THEOREM 1. Eigenvectors corresponding to distinct eigenvalues are linearly independent.

Proof. Let $\lambda_1, \lambda_2, \ldots, \lambda_m$ be a set of distinct eigenvalues of a matrix $A_{n,n}$ with $n \geqslant m$, and let $u^1, u^2, \ldots, u^m$ be their corresponding eigenvectors. Form the linear relation

$$c_1 u^1 + c_2 u^2 + \cdots + c_m u^m = 0$$

The problem becomes that of showing that

$$c_1 = c_2 = \cdots = c_m = 0$$

Pre-multiplying the linear relation first by A, then by A^2 etc., until A^{m-1}, we obtain

$$\begin{aligned}
&c_1 \lambda_1 u^1 + c_2 \lambda_2 u^2 + \cdots + c_m \lambda_m u^m = 0 \\
&c_1 \lambda_1^2 u^1 + c_2 \lambda_2^2 u^2 + \cdots + c_m \lambda_m^2 u^m = 0 \\
&\vdots \\
&c_1 \lambda_1^{m-1} u^1 + c_2 \lambda_2^{m-1} u^2 + \cdots + c_m \lambda_m^{m-1} u^m = 0
\end{aligned}$$

The above equations can be organized in the following form:

$$\left[c_1 u^1 \;\vdots\; c_2 u^2 \;\vdots\; \cdots\cdots \;\vdots\; c_m u^m \right] \begin{bmatrix} 1 & \lambda_1 & \cdots & \lambda_1^{m-1} \\ 1 & \lambda_2 & & \\ \vdots & \vdots & & \vdots \\ 1 & \lambda_m & & \lambda_m^{m-1} \end{bmatrix} = 0_{n,m}$$

The second matrix on the left-hand side is a Vandermonde matrix whose determinant is non-zero since the eigenvalues $\lambda_1, \ldots, \lambda_m$ are all distinct (see Exercise 2.1.15); hence

$$c_1 u^1 = c_2 u^2 = \cdots = c_m u^m = 0$$

However

$$u^1 \neq 0, \ldots, u^m \neq 0$$

Therefore

$$c_1 = c_2 = \cdots = c_m = 0$$

and the proof is complete.

In the second example the matrix A has multiple eigenvalues; however their corresponding eigenvectors were also found to be linearly independent. The reason is that A is real symmetric. The proof that a real symmetric matrix possesses linearly independent eigenvectors is on a different basis altogether, as will be shown further on.

In general one can find in one matrix the three cases combined; i.e. distinct eigenvalues with corresponding linearly independent eigenvectors, multiple eigenvalues with corresponding linearly independent eigenvectors and multiple eigenvalues with linearly dependent eigenvectors. The first two kinds are called *semi-*

simple eigenvalues and the last ones are called *non-semi-simple* eigenvalues. The following figure shows all possibilities:

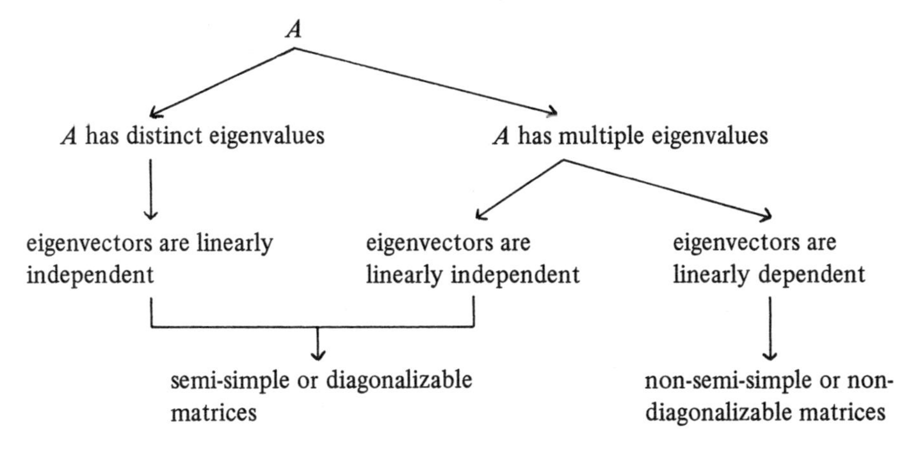

The question naturally arises how to know whether a given matrix is semi-simple or not. Unfortunately there is no easy answer in general, although there are some identifiable classes of matrix that are always semi-simple. But in general one must study the eigenvectors themselves. The reader can prove, using Exercise 4.1.14, that if

$$\text{rank}\,(\lambda_i I - A) = n - k_i, \quad i = 1, 2, \ldots, s$$

where n is the order of A, k_i is the order of multiplicity of λ_i and s is the number of distinct eigenvalues, then A is a semi-simple matrix. The following theorem concerns the general case.

THEOREM 2. $n - k_i \leqslant \text{rank}\,(\lambda_i I - A) \leqslant n - 1$

Proof. From the equations

$$(\lambda_i I - A)\, u^i = 0$$

we obtain, using the rank of the product of two matrices

$$\begin{aligned} 0 &\geqslant \rho(\lambda_i I - A) + \rho(u^i) - n \\ &= \rho(\lambda_i I - A) + 1 - n \end{aligned}$$

Therefore

$$\rho(\lambda_i I - A) \leqslant n - 1$$

To show that

$$\rho(\lambda_i I - A) \geqslant n - k_i$$

we assume the opposite of this proposition and show that it is false. If $\rho(\lambda_i I - A)$ were to be less than $n - k_i$, where k_i is the order of multiplicity of the eigenvalue

λ_i, this will result in a set of linearly independent eigenvectors of number greater than k_i. Applying this result to all other eigenvalues results in a number of linearly independent eigenvectors of number greater than n, which is impossible.

The reader may try to provide a counter-example to this theorem, by thinking of a situation in which A has two eigenvalues λ_1 and λ_2, where

$$\rho(\lambda_1 I - A) < n - k_1$$

and

$$\rho(\lambda_2 I - A) > n - k_2$$

such that the eigenvectors of λ_1 and λ_2 span the space $\mathbb{C}^{k_1+k_2=n}$. The reader can show that this cannot happen, by proceeding as follows: Let $u^1, u^2, \ldots, u^\sigma$ be the set of linearly independent eigenvectors corresponding to λ_1 and $u^{\sigma+1}, \ldots, u^{\sigma+\eta}$ be the set of linearly independent eigenvectors corresponding to λ_2, where of course $\sigma > k_1$ and $\eta < k_2$. From Exercise 4.1.14, one deduces that the vectors $u^1, u^2, \ldots, u^{\sigma+\eta}$ are linearly independent; choose $\tilde{u}$ such that

$$\begin{aligned} A\tilde{u}^1 &= \lambda_2 \tilde{u}^1 + u^{\sigma+\eta} \\ A\tilde{u}^2 &= \lambda_2 \tilde{u}^2 + \tilde{u}^1 \\ &\vdots \\ A\tilde{u}^{k_2-\eta} &= \lambda_2 \tilde{u}^{k_2-\eta} + \tilde{u}^{k_2-\eta-1} \end{aligned}$$

By so doing one can prove that $u^1, u^2, \ldots, u^\sigma, u^{\sigma+1}, \ldots, u^{\sigma+\eta}, \tilde{u}^1, \ldots, \tilde{u}^{k_2-\eta}$ are all linearly independent. If so, we obtain a number of linearly independent vectors greater than $k_1 + k_2$, which is impossible; and this completes the proof of the above Theorem 2.

Exercises 4.1

1. Find the eigenvalues and eigenvectors of the following matrices:

$$\begin{bmatrix} 1 & 2 & 0 \\ 0 & -1 & 0 \\ 0 & 0 & 3 \end{bmatrix}, \quad \begin{bmatrix} 1 & 2 & 0 \\ 2 & -1 & 0 \\ 0 & 0 & \sqrt{5} \end{bmatrix}, \quad \begin{bmatrix} 2 & 1 & 0 \\ 0 & 2 & 1 \\ 0 & 0 & 2 \end{bmatrix}$$

2. If A has the eigenvalues $\lambda_1, \ldots, \lambda_n$, show that A^m, with m integral, has the eigenvalues $\lambda_1^m, \ldots, \lambda_n^m$ and the same eigenvectors as A, only if A^m exists.
3. If A has an eigenvalue λ, show that $F(A) = a_0 I + a_1 A + a_2 A^2 + \cdots$ has the eigenvalue $f(\lambda) = a_0 + a_1\lambda + a_2\lambda^2 + \cdots$.
4. If A has the eigenvalue λ, show that cA (c is a scalar) has the eigenvalue $c\lambda$.
5. If u^i is an eigenvector of A, show that cu^i (c is a non-zero scalar) is also an eigenvector of A.
6. If $u^1, u^2, \ldots, u^m$ are eigenvectors of A corresponding to the same eigenvalue λ, show that $c_1 u^1 + c_2 u^2 + \cdots + c_m u^m$ is also an eigenvector corresponding to the same eigenvalue.
7. Show that the eigenvalues of a diagonal matrix are the diagonal elements; what are the eigenvectors? Discuss also the case of a triangular matrix.
8. If A is real, show that the eigenvalues appear in conjugate pairs; is this true of the eigenvectors?

9. If $L^{-1}L = LL^{-1} = I$, show that $A, L^{-1}AL, LAL^{-1}$ have the same eigenvalues. What about the eigenvectors? $A, L^{-1}AL$ and LAL^{-1} are called *similar* matrices or connected by a *similarity transformation.*
10. Show that if A has a zero eigenvalue, it is singular.
11. If A is nonsingular and has an eigenvalue λ, show that A^{-1}, where $A^{-1}A = AA^{-1} = I$, has an eigenvalue $1/\lambda$.
12. If U is unitary, show that A and U^*AU have the same eigenvalues.
13. If $A^n = I$, with n a positive integer, show that the eigenvalues of A are the nth root of unity. A is called an involutary matrix.
14. If $u^1, u^2, \ldots, u^\sigma$ are a set of linearly independent eigenvectors of A corresponding to an eigenvalue λ_1, and $u^{\delta+1}, \ldots, u^{\delta+\eta}$ are a set of linearly independent eigenvectors corresponding to an eigenvalue $\lambda_2 \neq \lambda_1$, show that $u^1, u^2, \ldots, u^{\delta+\eta}$ are all linearly independent.
15. Show that A and A^T have the same eigenvalues. Do these two matrices have the same eigenvectors?
16. If λ is a distinct eigenvalue of A, show that $\rho(\lambda I - A) = n - 1$.
17. If A is semi-simple, show that $\rho(A)$ = number of its non-zero eigenvalues. This theorem is useful for computing the rank of any matrix A. The method relies upon calculating the eigenvalues of A^*A. As A^*A is semi-simple, $\rho(A^*A) = \rho(A)$ = number of non-zero eigenvalues of A^*A. If $\rho(A) = r$, and the eigenvalues of A^*A are $\sigma_i^2 (i = 1, \ldots, n)$, where A is of order (m, n), then $\sigma_{r+1} = \sigma_{r+2} = \cdots = \sigma_n = 0$. $\sigma_i (i = 1, \ldots, n)$ are called the *singular values* of A. Usually the numerical methods test the singular values of A against a small tolerance. For further information, refer to Forsythe and Moler (1967), p. 5, and Wilkinson and Reinsch (1971), Contribution I/10.

18. Show that $\prod_{i=1}^{n} \lambda_i = |A|$ and $\sum_{i=1}^{n} \lambda_i = \text{tr } A$.

19. If $Au^i = \lambda_i u^i$, show that $\lambda_i = \langle u^i, Au^i \rangle / \langle u^i, u^i \rangle$. This expression is useful for calculating λ_i if u^i is known. Compare this method with that of calculating Au^i.
20. Show that AB and BA have the same eigenvalues.
21. If $AB = BA$, show that A and B have the same eigenvectors.
22. Find the different conditions on λ and b for which the equation $Au = \lambda u + b$ has a solution for u.
23. Find the condition on λ for which the set of linear equations

$$(A\lambda^2 + B\lambda + C)q = 0$$

has a non-trivial solution for the vector q. Hence obtain q, if

$$A = \begin{bmatrix} 1 & 0 \\ 0 & 0 \end{bmatrix}, \quad B = \begin{bmatrix} 1 & 0 \\ 0 & 1 \end{bmatrix}, \quad C = \begin{bmatrix} 1 & -1 \\ -1 & 1 \end{bmatrix}$$

24. Using the matrix

$$\begin{bmatrix} 0 & 1 & & & \\ & 0 & 1 & & \\ & & \ddots & \ddots & \\ & & & & 1 \\ -\dfrac{a_n}{a_0} & \cdots & & -\dfrac{a_2}{a_0} & -\dfrac{a_1}{a_0} \end{bmatrix}$$

show that any given polynomial $a_0\lambda^n + a_1\lambda^{n-1} + \cdots + a_{n-1}\lambda + a_n (a_0 \neq 0)$ of degree n may be regarded as the characteristic polynomial of a matrix of order n. This matrix is called the *companion* matrix of the given polynomial.

25. If A is any square complex matrix, show that there exists a unitary matrix U such that U^*AU is an upper triangular matrix with diagonal elements equal to the eigenvalues of A. This theorem is due to Schur. *Hint*: choose any nonsingular matrix Q_1 having in its first column the first eigenvector of A corresponding to λ_1; then use the Gram–Schmidt process to obtain from Q_1 a unitary matrix U_1 whose first column is proportional to that of Q_1. Thus

$$U_1^*AU_1 = \begin{bmatrix} \lambda_1 & B_1 \\ 0 & A_1 \end{bmatrix}$$

where A_1 is of order $n-1$ and has eigenvalues $\lambda_2, \ldots, \lambda_n$ (Why is this so?). Next form Q_2 having as first column the eigenvector of A_1 corresponding to λ_2, etc. After sufficient iterations, show how to construct U. Apply the theorem to

$$A = \begin{bmatrix} 5+5i & -1+i & -6-4i \\ -4-6i & 2-2i & 6+4i \\ 2+3i & -1+i & -3-2i \end{bmatrix}, \quad i^2 = -1.$$

26. If the largest magnitude of the eigenvalues of a matrix A is called the *spectral radius* $s(A) = \max_i |\lambda_i(A)|$, show that $s(A) \leqslant \|A\|$, i.e. that the eigenvalues of A cannot lie outside a circle of radius $\|A\|$. Should $\|A\|$ be subordinate to a vector norm?

27. Show that $\gamma(A) \geqslant \dfrac{\max|\lambda(A)|}{\min|\lambda(A)|}$. This formula provides a good approximation for $\gamma(A)$.

28. Prove Gershgorin's theorem, namely that the eigenvalues of A lie in the union of the circles $|\lambda - a_{ii}| = \sum_{j=1, j\neq i}^{n} |a_{ij}|$. *Hint*: write $Au = \lambda u$ in the form $(\lambda - a_{ii})u_i = \sum_{i=1, j\neq i}^{n} a_{ij}u_j$. Show how to use this theorem to prove that a diagonal dominant matrix is nonsingular.

29. Show that $s((I+L)^{-1}U) < 1$, where L and U are the lower and upper part of

$$A = \begin{bmatrix} 0 & \dfrac{a_{12}}{a_{11}} & \dfrac{a_{13}}{a_{11}} & \cdots & \dfrac{a_{1n}}{a_{11}} \\ \dfrac{a_{21}}{a_{22}} & 0 & \dfrac{a_{23}}{a_{22}} & & \\ \vdots & & & & \\ \dfrac{a_{n1}}{a_{nn}} & & \cdots\cdots\cdots & & 0 \end{bmatrix}$$

and where

$$\sum_{j \neq i} \left| \frac{a_{ij}}{a_{ii}} \right| < 1.$$

Hint: consider the equation

$$\left| (I + L) - \frac{U}{\lambda} \right| = 0, \text{ in } \lambda;$$

show that $|\lambda| \geqslant 1$ does not satisfy it, since $(I + L) - U/\lambda$ is diagonal dominant.

30. If A is nilpotent, i.e. $A^k = 0$ for some positive integer k, show that

(a) $|A| = 0$
(b) $\operatorname{tr} A = 0$
(c) $|I + A| = 1$

31. If A is idempotent, i.e. $A^k = A$ for some positive integer k, show that, if $k = 2$,

(a) $|A| = 0$ or unity
(b) $\lambda(A)$ is either zero or unity
(c) $\rho(A) = \operatorname{tr} A$

32. Show that if A is semi-simple, its eigenvalues and eigenvectors define the matrix uniquely. *Hint*: let A and A' be such matrices; to show that $A = A'$ is to show that $(A - A')x = 0, \forall x$.

33. If $s(A) < 1$ for any matrix A, show that $\lim_{k \to \infty} A^k = 0$. *Hint*: show that $A^k x = 0, \forall x$. Expand x in terms of the eigenvectors of A. If A is non-semi-simple, choose a new independent basis as in Theorem 2 of Section 4.1.

34. If $A = \text{quasidiag}(A_1, A_2, \ldots, A_m)$, obtain its eigenvectors in relation to the eigenvectors of $A_1, \ldots, A_m$. Show that if $A_1, \ldots, A_m$ are semi-simple matrices, A is semi-simple. Should this result hold true only if $A_1, \ldots, A_m$ have no eigenvalues in common?

35. Let A, B, C and X be matrices of the second order. Show that every eigenvalue of a solution X of $AX^2 + BX + C = 0$ is a root of $|A\lambda^2 + B\lambda + C| = 0$. Show how to obtain X, and apply the result on an example.

4.2. Computation of the characteristic polynomial

This section supplies the reader with various numerical methods for calculating the characteristic polynomial of a square matrix A. These methods are very much simpler than expansion of $|\lambda I - A|$. Besides, if the order of A is high, it becomes impracticable to evaluate this function using the known methods of determinant expansion.

A simple algorithm often used to obtain the characteristic polynomial is the one due to Leverrier, which relies mainly on the theory of equations. Let the characteristic polynomial of A be given by

$$|\lambda I - A| = \lambda^n + a_{n-1}\lambda^{n-1} + \cdots + a_0;$$

where it is required to calculate the coefficients $a_{n-1}, a_{n-2}, \ldots, a_0$. Define the matrices

$$
\begin{aligned}
B_{n-1} &= I_n \\
B_{n-2} &= a_{n-1} I_n + A B_{n-1} \\
B_{n-3} &= a_{n-2} I_n + A B_{n-2} \\
&\vdots \\
B_1 &= a_2 I_n + A B_2 \\
B_0 &= a_1 I_n + A B_1
\end{aligned}
$$

Then

$$
\begin{aligned}
a_{n-1} &= -\operatorname{tr}(A B_{n-1}) \\
a_{n-2} &= -\tfrac{1}{2} \operatorname{tr}(A B_{n-2}) \\
a_{n-3} &= -\tfrac{1}{3} \operatorname{tr}(A B_{n-3}) \\
&\vdots \\
a_1 &= -\frac{1}{n-1} \operatorname{tr}(A B_1) \\
a_0 &= -\frac{1}{n} \operatorname{tr}(A B_0)
\end{aligned}
$$

Proof. Define σ_r such that

$$\sigma_r = \sum_{i=1}^{n} \lambda_i^r$$

where λ_i is a root of the characteristic equation. Then from the theory of equations, we obtain

$$\sigma_k + a_{n-1}\sigma_{k-1} + \cdots + a_{n-k+1}\sigma_1 + k a_{n-k} = 0, \quad k = 1, \ldots, n$$

A proof of the above relation is given in Appendix 2. Also, from Exercises 4.1.2 and 4.1.18, one can show that

$$\operatorname{tr} A^r = \sum_{i=1}^{n} \lambda_i^r = \sigma_r$$

Therefore

$$\operatorname{tr} A^k + a_{n-1} \operatorname{tr} A^{k-1} + \cdots + a_{n-k+1} \operatorname{tr} A + k a_{n-k} = 0, \quad k = 1, \ldots, n.$$

Substituting $k = 1, \ldots, n$ we obtain, using the identities defining the B matrices, the coefficients of the characteristic polynomial in the above given form.

Example. Obtain the characteristic polynomial of

$$A = \begin{bmatrix} 5 & -2 & -4 \\ -2 & 2 & 2 \\ -4 & 2 & 5 \end{bmatrix}$$

It is given by

$$|\lambda I - A| = \lambda^3 + a_2\lambda^2 + a_1\lambda + a_0$$

where

$$a_2 = -\text{tr}\, A = -12$$

The matrix B_1 is given by

$$B_1 = -12I + A = \begin{bmatrix} -7 & -2 & -4 \\ -2 & -10 & 2 \\ -4 & 2 & -7 \end{bmatrix}, \quad \text{and } AB_1 = \begin{bmatrix} -15 & 2 & -4 \\ 2 & -12 & -2 \\ 4 & -2 & -15 \end{bmatrix}$$

Hence

$$a_1 = -\tfrac{1}{2}\, \text{tr}(AB_1) = 21$$

Also

$$B_0 = a_1 I + AB_1 = \begin{bmatrix} 6 & 2 & 4 \\ 2 & 9 & -2 \\ 4 & -2 & 6 \end{bmatrix}$$

from which we obtain

$$AB_0 = 10I$$

giving

$$a_0 = -\frac{1}{3}\, \text{tr}\,(AB_0) = -10$$

The characteristic polynomial is therefore

$$\lambda^3 - 12\lambda^2 + 21\lambda - 10.$$

Leverrier's method, although the oldest, is a very accurate one for computing the characteristic polynomial. It often serves for calculating the determinant of a matrix. The latter is calculated from

$$|\lambda I - A|_{\lambda=0} = a_0$$

The only drawback of Leverrier's algorithm becomes pronounced when A is large, since it becomes slow in comparison with other methods. Of the latter, we recall the most popular known as Krylov's method. It is by far much faster than Leverrier's, but it suffers from round-off errors for it relies on solving a set of linear equations. For a brief discussion of the method the reader should refer to

Exercise 4.10.49. An interesting method that we should mention is the one devised by Danilevsky. The connection between a given polynomial and its companion matrix (see Exercise 4.1.24) inspired him to devise an algorithm for computing the characteristic polynomial of A. His method relies upon constructing a sequence of similarity transformations which transform A into a companion matrix. Writing both matrices as follows

$$A = \begin{bmatrix} a_{11} & a_{12} & \cdots & a_{1n} \\ a_{21} & a_{22} & \cdots & a_{2n} \\ \vdots & & & \vdots \\ a_{n1} & & \cdots & a_{nn} \end{bmatrix}, \quad C = \begin{bmatrix} 0 & 1 & 0 & \cdots & 0 \\ 0 & 0 & 1 & \cdots & 0 \\ \vdots & & \ddots & \ddots & \\ & & & 0 & 1 \\ p_n & & \cdots & p_2 & p_1 \end{bmatrix}$$

Danilevsky showed that we can find $(n - 1)$ similarity transformations, the successive performance of which will realize the desired transition. To obtain the first transformation, divide the first row of A by a_{12} and then by inducing a set of elementary column operations, cancel all other elements in the first row. The first elementary column operation P_1 is then given by

$$P_1 = \begin{bmatrix} 1 & 0 & 0 & \cdots & 0 \\ \dfrac{-a_{11}}{a_{12}} & \dfrac{1}{a_{12}} & \dfrac{-a_{13}}{a_{12}} & \cdots & \dfrac{-a_{1n}}{a_{12}} \\ 0 & 0 & 1 & & 0 \\ \vdots & & & & \vdots \\ 0 & 0 & & \cdots & 1 \end{bmatrix},$$

$$P_1^{-1} = \begin{bmatrix} 1 & 0 & 0 & \cdots & 0 \\ a_{11} & a_{12} & a_{13} & \cdots & a_{1n} \\ 0 & 0 & 1 & \cdots & 0 \\ \vdots & & & & \vdots \\ 0 & & \cdots & & 1 \end{bmatrix}$$

Note that multiplication of AP_1 from the left by P_1^{-1} does not change the transformed row. And the process can be repeated to obtain $P_2, \ldots, P_{n-1}$. Danilevsky's method is also useful in obtaining the eigenvectors of A. If y is an eigenvector of C, then $x = P_1 \ldots P_{n-1} y$ is an eigenvector of A (see Exercise 4.1.9). For more computational information, the reader may refer to Faddeeva (1959), p. 166.

A comparative examination of the above methods reveals that Danilevsky's method has probably the least number of operations. It also secures a comparatively high accuracy, since the calculation of the P matrices and the P^{-1} matrices is trivial. Moreover, the method makes possible the determination of the eigenvectors in an easy way.

4.3. Computation of eigenvalues and eigenvectors

Methods based on finding the roots of the characteristic polynomial to represent the eigenvalues of a matrix are unreliable, since the coefficients of the polynomial have to be determined in rounded arithmetic, and the polynomial's roots are sensitive in general to small perturbations in the coefficients. We should also mention that solving a high-order algebraic equation is not easy. This is why numerical analysts have devised algorithms for computing the eigenvalues and eigenvectors of a matrix and bypassing the characteristic polynomial. Of these algorithms we mention the most famous, called the *power method of Von Mises.* It determines the eigenvalues one by one, starting by the largest in magnitude. The method becomes tremendously important when one needs only to calculate the eigenvalue of maximum modulus, as in some applications of economic theory. In demonstrating the power method, we will consider only the case when A is semi-simple, and leave the case when A is non-semi-simple as an exercise for the reader, which he will be able to tackle once he is acquainted with the eigenvalue problem of non-semi-simple matrices.

Let $\lambda_1, \lambda_2, \ldots, \lambda_n$ be the eigenvalues of a semi-simple matrix A with corresponding eigenvectors $u^1, u^2, \ldots, u^n$. Assume without loss of generality that

$$|\lambda_1| \geqslant |\lambda_2| \geqslant \cdots \geqslant |\lambda_n|$$

We start by calculating λ_1. Choose any vector $x \neq 0$, and expand x in terms of the eigenvectors of A as follows:

$$x = \sum_{i=1}^{n} c_i u^i$$

Pre-multiplying the above expansion by $A, A^2, \ldots$, we obtain:

$$x^1 = Ax = \sum_{i=1}^{n} c_i \lambda_i u^i$$

$$x^2 = Ax^1 = \sum_{i=1}^{n} c_i \lambda_i^2 u^i$$

$$\vdots$$

$$x^m = Ax^{m-1} = \sum_{i=1}^{n} c_i \lambda_i^m u^i$$

And if m is large enough, we notice that by iteration x^m tends to $\lambda_1^m c_1 u^1$ since

$$x^m = \sum_{i=1}^{n} c_i \lambda_i^m u^i = \lambda_1^m \left(c_1 u^1 + \sum_{i=2}^{n} \left(\frac{\lambda_i}{\lambda_1} \right)^m c_i u^i \right)$$

Therefore λ_1 is obtained from two successive iterations, by dividing the corresponding elements of x^m and x^{m-1}. As for u^1, it is directly proportional to x^m.

Example. Obtain λ_1 if

$$A = \begin{bmatrix} 5 & -2 & -4 \\ -2 & 2 & 2 \\ -4 & 2 & 5 \end{bmatrix}$$

Choose $x^T = (1 \quad 0 \quad 0)$, and obtain

$$x^1 = Ax = \begin{bmatrix} 5 \\ -2 \\ -4 \end{bmatrix}$$

$$x^2 = Ax^1 = \begin{bmatrix} 45 \\ -22 \\ -44 \end{bmatrix}$$

$$x^3 = Ax^2 = \begin{bmatrix} 445 \\ -222 \\ -444 \end{bmatrix}$$

We notice that for large m, the iterative method gives

$$\lambda_1 \approx \frac{445}{45} \approx \frac{-222}{-22} \approx \frac{-444}{-44} \approx 10$$

The eigenvector u^1, proportional to x^m, is found to be

$$u^1 = \begin{bmatrix} 2 \\ -1 \\ -2 \end{bmatrix}$$

To obtain λ_2, we assume that v^1 is the known reciprocal vector of u^1 and apply the power method on the new matrix A_1 given by

$$A_1 = A - \lambda_1 u^1 v^{1*}$$

The reader can prove that A_1 has the same eigenvalues as A, except that λ_1 is transferred to the origin. It also has the same eigenvectors as A. Hence it becomes obvious that we can reapply the power method to A_1 having λ_2 as its eigenvalue with largest magnitude.

The reader may inquire whether one can determine v^1. Unfortunately there is no easy way of doing so. But if A is real symmetric, then $v^1 = u^1$, a result which can be established once the reader has studied the eigenvalue problem of real symmetric and Hermitian matrices in the next sections. In this case the power method can be applied continuously. Such a procedure for modifying the eigenvalue problem is called *deflation.* A comprehensive description of the available deflation methods can be found in Wilkinson (1965), Chapter 9.

If A is not symmetric, having known only λ_1 and u^1, we can forget about v^1, and proceed directly with the Schur algorithm; see Exercise 4.1.25. The matrix A_1 appearing in the algorithm of Schur is of order $n - 1$ and has the eigenvalues $\lambda_2, \ldots, \lambda_n$. Although its eigenvector is shorter in size than that of A, there is a relation between them. Therefore applying the power method on A_1 gives λ_2 and its corresponding eigenvector y^2, which is indispensable for the continuation of the Schur algorithm. And by scaling correctly the equations

$$\begin{bmatrix} \lambda_1 & B_1 \\ 0 & A_1 \end{bmatrix} \begin{bmatrix} 1 \\ y^2 \end{bmatrix} = \lambda_2 \begin{bmatrix} 1 \\ y^2 \end{bmatrix}$$

we obtain the eigenvector of $U_1^* A U_1$ and upon pre-multiplying it by U_1 we obtain u^2, etc. . . .

For practical computation the power method is only conditionally useful, since convergence relies upon the factors $|\lambda_i/\lambda_1|$. If these factors are not sufficiently small, slow convergence is often met. In obtaining λ_2, the factors affecting convergence are $|\lambda_i/\lambda_2|$. This is why the power method converges slowly if the moduli of the eigenvalues are not well separated. The choice of the initial vector x is also important. Not knowing beforehand the eigenvectors, we might blindly choose x, such that c_1 is equal to zero. The power method will then by iteration give λ_2 and its corresponding eigenvector. It is only thanks to round-off errors, that it may eventually give λ_1 and its corresponding eigenvector. One other main disadvantage of the method is that it computes only one eigenvalue and one eigenvector and we have to resort to other additional algorithms.

A refinement of the power method called the *inverse iteration method* of Wielandt is often used to calculate accurately the eigenvectors when the eigenvalues are not well separated; for in this case, the power method converges slowly. For information about the inverse iteration method, refer to Stoer and Bulirsch (1980), p. 356; see also Section 4.10.

Perhaps the *LR* and *QR* algorithms are the best currently used methods for the numerical solution of eigenvalue problems. In the *LR* algorithm of Rutishauser (1958), beginning with the matrix A, one generates a sequence of matrices A_i which converge to an upper triangular matrix whose diagonal elements are the eigenvalues of A. The A_i matrices are obtained during the Gauss elimination method of decomposition. Starting with A, compute a lower triangular matrix L with $l_{ii} = 1$ and an upper triangular matrix R, such that $A = LR$; see Section 3.9a.

By forming the matrix sequence

$$A_1 = RL = L_1 R_1$$

$$A_2 = R_1 L_1 = L_2 R_2$$

$$\vdots$$

$$A_m = R_{m-1} L_{m-1} = L_m R_m,$$

it will be found after sufficient iterations that L_m tends to a unit matrix; see Stoer and Bulirsch (1980), p. 365. R_m will then contain in its diagonal the eigenvalues of A. This is because A and A_m are similar, since for two successive iterations

$$A_i = L_i R_i = L_i A_{i+1} L_i^{-1}$$

One main advantage of the *LR* algorithm over previous methods is its simplicity. It relies on the Gauss elimination method for decomposition. Perhaps it suffers, like most methods, when the eigenvalues are not well separated. It has also one major drawback, that is when a triangular decomposition for the A_i matrices does not exist. In this case, the algorithm breaks down. But this is not rare in occurrence, if A_i has one zero leading principal minor. Problems of accuracy are also often met if A_i has poor pivots. Briefly it suffers from the same symptoms which we encounter when running a Gauss elimination scheme without using a permutation matrix. Therefore the *LR* algorithm is more efficient when the matrix is positive definite; see Cholesky decomposition in Section 3.9a. For further reading about the *LR* algorithm, and computational difficulties, the reader is referred to Wilkinson (1965), pp. 498, 538.

The difficulties associated with the *LR* algorithm led Francis (1961, 1962) and Kublanovskaya (1961) to devise a modification which replaces L by a unitary matrix Q. Such a decomposition always exists for a nonsingular matrix A (see Young and Gregory (1973), p. 921) and it brings great improvements over the *LR* algorithm. The iterations will similarly read

$$A_i = R_{i-1} Q_{i-1} = Q_i R_i = Q_{i-1}^{-1} A_{i-1} Q_{i-1}$$

The *LR* and *QR* algorithms are well implemented practically for *tridiagonal matrices* (matrices having only three main diagonals), *Hessenberg matrices* (almost triangular form with one more diagonal beside the main diagonal) and *band matrices* (matrices having few main diagonals). The reason arises from the huge number of arithmetical operations. Stoer and Bulisch (1980), p. 368, report that the *LR* algorithm requires for every iteration, if the matrix is dense, $2/3\ n^3$ operations. This is a problem which can be met especially in large matrices. Usually the matrix is first transformed to one of the above kinds, before applying the *LR* or *QR* algorithms in order to save computation. Wilkinson (1965), pp. 347–355, discusses various methods for reducing a matrix to Hessenberg form. It is shown that the *QR* algorithm preserves the Hessenberg form or the tridiagonal form when A is Hermitian or real symmetric.

It appears that the methods employed to solve the eigenvalue problem depend on many factors. These factors include whether A is symmetric, whether any eigenvectors are required and how many eigenvalues are to be determined. There is no such method which can be called ideal for all cases. Usually there is a choice of alternatives. Even among these alternatives, methods differ in accuracy, time spending and computational efficiency. For large matrices, we recommend the power method, especially if a few of the largest eigenvalues and corresponding eigenvectors are all that are required. Perhaps the inverse iteration method is the most powerful and accurate for computing eigenvectors; see Wilkinson (1965), p. 622. A general method for computing all eigenvalues is the *QR* algorithm. Especially when A is Hermitian or real symmetric, the *QR* algorithm preserves the Hessenberg form of A. But these matrices have already well-conditioned eigenproblems (see Section 4.10), and so many algorithms can be applied to them; let alone positive definite matrices appearing in the formulation of many dynamical systems. An example is the generalized eigenvalue problem

$$Ax = \lambda Bx$$

where usually A and B are symmetric and B is positive definite. B is usually decomposed using the Cholesky algorithm (see Section 3.9a) in the form

$$B = R^T R$$

The eigenvalue problem will then read

$$Ax = \lambda R^T R x$$

And if we choose a vector y such that

$$y = Rx$$

we obtain finally

$$R^{T-1} A R^{-1} y = Cy = \lambda y$$

which is an eigenvalue problem of the matrix C. For a procedure for forming C, see Wilkinson and Reinsch (1971), Contribution II/10.

If B is only positive semi-definite (having some zero eigenvalues), the above transformation is modified; see Peters and Wilkinson (1970). If on the other hand A and B are any square matrices, the reader is referred to the *Qz* algorithm described by Moler and Stewart (1971).

4.4. Diagonalization of semi-simple matrices

Going back to the eigenvalue problem, we can write it for all eigenvalues and eigenvectors in the following form

$$\begin{aligned} Au^1 &= \lambda_1 u^1 \\ Au^2 &= \lambda_2 u^2 \\ &\vdots \\ Au^n &= \lambda_n u^n \end{aligned}$$

The above equations can be similarly organized in a more concise form as follows:

$$A\begin{bmatrix} u^1 u^2 & \ldots & u^n \end{bmatrix} = \begin{bmatrix} u^1 u^2 & \ldots & u^n \end{bmatrix}\begin{bmatrix} \lambda_1 & & & \\ & \lambda_2 & & \\ & & \ddots & \\ & & & \lambda_n \end{bmatrix}$$

i.e.

$$AT = TD_\lambda$$

where the matrix T whose columns are the eigenvectors of A, is called the *modal matrix* for A and where D_λ is a diagonal matrix containing the eigenvalues.

Although the above matrix equation is valid for any matrix A, it takes a better and more useful form when A is semi-simple, since, in this case, the eigenvectors are linearly independent, which makes T nonsingular. Consequently there exists a matrix T^{-1} such that $T^{-1}T = I$, and the equation reads

$$T^{-1}AT = D_\lambda$$

The above result is very important, announcing that, if A is semi-simple, it can be diagonalized by a similarity transformation.

4.5. The eigenvalue problem for Hermitian and real symmetric matrices

Because Hermitian and real symmetric matrices are frequently met in physical problems, they deserve a proper treatment of their own. Before we proceed we establish the following theorem.

THEOREM 1. The eigenvalues of a real symmetric or Hermitian matrix are real, and two eigenvectors corresponding to two distinct eigenvalues are orthogonal.

Proof. Let the eigenvalue problem for two distinct eigenvalues be written as follows:

$$Au^i = \lambda_i u^i$$
$$Au^j = \lambda_j u^j$$

Taking the conjugate transpose of the second equation, post-multiplying it by u_i, and pre-multiplying the first equation by u^{j*}, we obtain

$$u^{j*}Au^i = \lambda_i u^{j*}u^i$$
$$u^{j*}Au^i = \bar{\lambda}_j u^{j*}u^i$$

Subtracting we obtain

$$0 = (\lambda_i - \bar{\lambda}_j)\langle u^j, u^i\rangle$$

For $i = j$ we get

$$0 = (\lambda_i - \bar{\lambda}_i)\langle u^i, u^i\rangle$$

But as

$$\langle u^i, u^i\rangle \neq 0$$

unless

$$u^i = 0$$

which is not true, since u^i is an eigenvector of A—in other words a non-trivial solution of the eigenvalue problem. Hence we obtain

$$\lambda_i = \bar{\lambda}_i$$

And for $i \neq j$, we get

$$0 = (\lambda_i - \lambda_j)\langle u^j, u^i\rangle$$

showing when

$$\lambda_i \neq \lambda_j$$

that

$$\langle u^j, u^i\rangle = 0$$

and the proof is complete.

From the above proof the reader will notice, with multiple eigenvalues, i.e. when $\lambda_i = \lambda_j$, although $i \neq j$, that u^j and u^i may or may not be orthogonal. To show this in an example, consider the eigenvalue problem for the two following matrices

$$\begin{bmatrix} 5 & -2 & -4 \\ -2 & 2 & 2 \\ -4 & 2 & 5 \end{bmatrix}, \begin{bmatrix} 1 & 2 & 0 \\ 2 & -1 & 0 \\ 0 & 0 & \sqrt{5} \end{bmatrix}$$

For the first matrix, $\lambda = 1$ is an eigenvalue of multiplicity 2 with two non-orthogonal eigenvectors. Whereas for the second matrix, $\lambda = \sqrt{5}$ is an eigenvalue of multiplicity 2 with two orthogonal eigenvectors.

Now we are in a position to explain the method followed to diagonalize A. If A has distinct eigenvalues, the eigenvectors are all orthogonal in pairs as a result of the above theorem. Moreover if each eigenvector is normalized such that its norm is unity, the normalized modal matrix is unitary and we obtain

$$T^{-1} = T^*$$

This shows that Hermitian or real symmetric matrices can be diagonalized by a congruent (conjunctive) transformation as follows:

$$T^*AT = D_\lambda$$

T is real when A is real symmetric (the reader should be able to see why this is so).

Example. Rotate the axis x, y to x', y' such that the equation of the ellipsoid

$$13x^2 + 13y^2 + 10xy = 72$$

is transformed to its canonical form.
The equation can be written as

$$[x \quad y]\begin{bmatrix} 13 & 5 \\ 5 & 13 \end{bmatrix}\begin{bmatrix} x \\ y \end{bmatrix} = 72$$

Let the relation between the old and new coordinates be represented by

$$\begin{bmatrix} x \\ y \end{bmatrix} = T\begin{bmatrix} x' \\ y' \end{bmatrix}$$

Then, substituting in the equation of the ellipsoid, we obtain

$$[x' \quad y']\ T^T\begin{bmatrix} 13 & 5 \\ 5 & 13 \end{bmatrix}T\begin{bmatrix} x' \\ y' \end{bmatrix} = 72$$

So for the ellipsoid equation to be written in the canonical form we must have

$$T^T\begin{bmatrix} 13 & 5 \\ 5 & 13 \end{bmatrix}T$$

diagonalized. This is feasible if T is the modal matrix of the real symmetric matrix

$$\begin{bmatrix} 13 & 5 \\ 5 & 13 \end{bmatrix}$$

Solving we obtain

$$\lambda_1 = 8, \quad \lambda_2 = 18$$

and

$$u^1 = \begin{bmatrix} 1 \\ -1 \end{bmatrix}, \quad u^2 = \begin{bmatrix} 1 \\ 1 \end{bmatrix}$$

From this we obtain the normalized modal matrix

$$T = \begin{bmatrix} \frac{1}{\sqrt{2}} & \frac{1}{\sqrt{2}} \\ -\frac{1}{\sqrt{2}} & \frac{1}{\sqrt{2}} \end{bmatrix}$$

and the new equation of the ellipsoid in its canonical form becomes

$$8x'^2 + 18y'^2 = 72$$

The reader can use the above method to discover whether the equation

$$ax^2 + by^2 + cz^2 + dxy + exz + fyz + gx + hy + jz = 1$$

represents a hyperboloid, an ellipsoid or paraboloid. In general if the eigenvalues of

$$\begin{bmatrix} a & d/2 & e/2 \\ d/2 & b & f/2 \\ e/2 & f/2 & c \end{bmatrix}$$

are all positive, the equation represents an ellipsoid; if some are positive and others negative, it represents a hyperboloid; and if some are zeros, the equation represents a paraboloid. When does it represent a sphere?

In general, one can construct a unitary matrix T which diagonalizes A through a congruent transformation. If A has distinct eigenvalues, we proceed as above. The above methods can also be applied when A has multiple eigenvalues, but with corresponding orthogonal eigenvectors. If on the other hand A has multiple eigenvalues with corresponding non-orthogonal eigenvectors, we proceed as follows. Let λ be an eigenvalue of multiplicity m; then the eigenvalues with their corresponding eigenvectors can be ordered as

$$\lambda, \lambda, \ldots, \lambda, \lambda_{m+1}, \ldots, \lambda_n$$

$$u^1, u^2, \ldots, u^m, u^{m+1}, \ldots, u^n$$

Now the vectors $u^{m+1}, \ldots, u^n$ are orthogonal to each other and to the rest as a result of the above theorem. What is left is to choose a new set of orthogonal vectors $u'^1, u'^2, \ldots, u'^m$ each being orthogonal to $u^{m+1}, \ldots, u^n$ together with each being an eigenvector of A. The procedure we shall use is the Gram–Schmidt algorithm explained in chapter three.

Construction: Let $u'^1 = u^1$
Let $u'^2 = u^2 + \alpha u^1$

Now u'^2 is an eigenvector of A, for it is a combination of eigenvectors corresponding to the same eigenvalue λ. Also u'^2 is orthogonal to $u^{m+1}, \ldots, u^n$ since the latter are orthogonal to u^1 and u^2. What remains is to make u'^2 orthogonal to u'^1 i.e. to u^1. We obtain using the Gram–Schmidt process explained before:

$$\alpha = -\frac{\langle u^1, u^2 \rangle}{\langle u^1, u^1 \rangle}$$

Next we choose

$$u'^3 = u^3 + \alpha u^1 + \beta u^2$$

Using the same reasoning, we obtain as explained before:

$$\begin{bmatrix} \langle u^1, u^1 \rangle & \langle u^1, u^2 \rangle \\ \langle u^2, u^1 \rangle & \langle u^2, u^2 \rangle \end{bmatrix} \begin{bmatrix} \alpha \\ \beta \end{bmatrix} = - \begin{bmatrix} \langle u^1, u^3 \rangle \\ \langle u^2, u^3 \rangle \end{bmatrix}$$

and the process can be repeated until we obtain u'^m. Note that the Grammian matrix of the above equations is nonsingular, since the eigenvectors of a Hermitian or a real symmetric matrix are linearly independent. The latter is a special result of a theorem which we are going to prove shortly.

Example. Obtain T for

$$A = \begin{bmatrix} 5 & -2 & -4 \\ -2 & 2 & 2 \\ -4 & 2 & 5 \end{bmatrix}$$

Solving the characteristic equation, we obtain

$$\lambda_1 = 1, \quad \lambda_2 = 1, \quad \lambda_3 = 10$$

$$u^1 = \begin{bmatrix} -1 \\ -2 \\ 0 \end{bmatrix}, \quad u^2 = \begin{bmatrix} -1 \\ 0 \\ -1 \end{bmatrix}, \quad u^3 = \begin{bmatrix} 2 \\ -1 \\ -2 \end{bmatrix}$$

Choose $u'^1 = u^1, u'^2 = u^2 + \alpha u^1$, such that

$$\alpha = -\frac{\langle u^1, u^2 \rangle}{\langle u^1, u^1 \rangle} = -\frac{1}{5}$$

Hence

$$u'^{2^T} = [-4/5 \quad 2/5 \quad -1]$$

and the matrix T becomes

$$T = \begin{bmatrix} \frac{-1}{\sqrt{5}} & \frac{-4}{3\sqrt{5}} & \frac{2}{3} \\ \frac{-2}{\sqrt{5}} & \frac{2}{3\sqrt{5}} & \frac{-1}{3} \\ 0 & \frac{-5}{3\sqrt{5}} & \frac{-2}{3} \end{bmatrix}$$

DEFINITION. A matrix A is called *normal* if $AA^* = A^*A$.

THEOREM 2. A square matrix A is unitarily diagonalized by a congruent transformation, if and only if A is normal.

Proof. Using Exercise 4.1.25, we have

$$U^*AU = \begin{bmatrix} \lambda_1 & b_{12} & & \cdots & b_{1n} \\ & \lambda_2 & b_{23} & \cdots & b_{2n} \\ & & & & \vdots \\ & & 0 & & \lambda_n \end{bmatrix} = B$$

Suppose A is normal, then B is also normal, as the reader can verify. Equating elements of B^*B and BB^*, we obtain for the element (1, 1)

$$\lambda_1\bar{\lambda}_1 = \lambda_1\bar{\lambda}_1 + b_{12}\bar{b}_{12} + b_{13}\bar{b}_{13} + \cdots + b_{1n}\bar{b}_{1n}$$

which gives

$$\sum_{i=2}^{n} |b_{1i}|^2 = 0$$

from which we conclude that $b_{1i} = 0$, for $i = 2, \ldots, n$; the same reasoning can be applied again to the element (2, 2), and so on. We thus obtain $B = \text{diag}(\lambda_1, \lambda_2, \ldots, \lambda_n)$. And conversely if U^*AU is diagonal, A is normal, which completes the proof.

One direct and important application of this theorem is that real symmetric, Hermitian, real skew-symmetric, skew-Hermitian and unitary matrices being all normal can be then unitarily diagonalized by a congruent transformation. In other words all these matrices are semi-simple.

4.6. Positive definite matrices

Positive definite matrices appear in many applications. We frequently meet them in optimization, solution of engineering systems and in problems related to convergence criteria.

DEFINITION. A Hermitian matrix A is called *positive definite*, if for all vectors $x \neq 0$, $x^*Ax > 0$.

Example. Find the condition on the impedance matrix Z of an n-port electrical network that it represents a system of R^+, L^+ and C^+.

The condition that a network contains positive resistances, positive inductances and positive capacitances is that the active power dissipated in the network should be greater than zero, i.e.

$$\text{Re } v^*i > 0$$

Substituting from

$$v = Zi$$

we obtain

$$\text{Re } i^*Z^*i > 0$$

i.e.

$$\tfrac{1}{2}i^*(Z^* + Z)i > 0, \quad \forall i \neq 0$$

in other words, the condition is that the Hermitian part of Z should be positive definite.

A is called a *negative definite* matrix if $x^*Ax < 0$, $\forall x \neq 0$. Also if $x^*Ax \geqslant 0$, $\forall x \neq 0$, A is called a *positive semidefinite* matrix; whereas if $x^*Ax \leqslant 0$, $\forall x \neq 0$, then A is called a *negative semidefinite* matrix.

THEOREM 1. A necessary and sufficient condition that A is positive definite is that any one of the following conditions is satisfied:

(i) All leading principal minors are positive.
(ii) The eigenvalues of A are all positive.
(iii) A can be put in the form $A = QQ^*$, where Q is a nonsingular matrix.

Proof. We prove only the necessity – sufficiency is left to the reader. Write x^*Ax as follows:

$$[\bar{x}_1 \quad \bar{x}_2 \quad \dots \quad \bar{x}_n] \begin{bmatrix} a_{11} & a_{12} & \dots & a_{1n} \\ \bar{a}_{12} & a_{22} & & \\ \vdots & & & \\ \bar{a}_{1n} & & \dots & a_{nn} \end{bmatrix} \begin{bmatrix} x_1 \\ x_2 \\ \\ x_n \end{bmatrix} > 0, \quad \forall x \neq 0$$

Choose $x_1 \neq 0$ and $x_2 = \cdots = x_n = 0$. This gives

$$|x_1|^2 a_{11} > 0$$

from which we obtain

$$a_{11} > 0$$

Choose $x_1, x_2 \neq 0$ and $x_3 = \cdots = x_n = 0$. This gives

$$|x_1|^2 a_{11} + |x_2|^2 a_{22} + \bar{x}_1 a_{12} x_2 + x_1 \bar{a}_{12} \bar{x}_2 > 0$$

i.e.

$$a_{11}\left(x_1 + x_2 \frac{a_{12}}{a_{11}}\right)\left(\bar{x}_1 + \bar{x}_2 \frac{\bar{a}_{12}}{a_{11}}\right) + \left(a_{22} - \frac{a_{12}\bar{a}_{12}}{a_{11}}\right)|x_2|^2 > 0$$

or that

$$a_{11}\left|x_1 + x_2 \frac{a_{12}}{a_{11}}\right|^2 + \left(a_{22} - \frac{a_{12}\bar{a}_{12}}{a_{11}}\right)|x_2|^2 > 0$$

From which the reader can check that

$$\begin{vmatrix} a_{11} & a_{12} \\ \bar{a}_{12} & a_{22} \end{vmatrix} > 0$$

Next choose $x_1, x_2, x_3 \neq 0$ and $x_4 = \cdots = x_n = 0$. This gives

$$a_{11}|x_1|^2 + a_{22}|x_2|^2 + a_{33}|x_3|^2 + x_1\bar{x}_2\bar{a}_{12} + \bar{x}_1 x_2 a_{12} + x_1\bar{x}_3\bar{a}_{13}$$
$$+ \bar{x}_1 x_3 a_{13} + x_2\bar{x}_3\bar{a}_{23} + \bar{x}_2 x_3 a_{23} > 0$$

i.e.

$$a_{11}\left(x_1 + x_2\frac{a_{12}}{a_{11}} + x_3\frac{a_{13}}{a_{11}}\right)\left(\bar{x}_1 + \bar{x}_2\frac{\bar{a}_{12}}{a_{11}} + \bar{x}_3\frac{\bar{a}_{13}}{a_{11}}\right) + \left(a_{22} - \frac{a_{12}\bar{a}_{12}}{a_{11}}\right)|x_2|^2 +$$
$$+ \left(a_{23} - \frac{\bar{a}_{12}a_{13}}{a_{11}}\right)\bar{x}_2 x_3 + \left(\bar{a}_{23} - \frac{a_{12}\bar{a}_{13}}{a_{11}}\right)x_2\bar{x}_3 + \left(a_{33} - \frac{a_{13}\bar{a}_{13}}{a_{11}}\right)|x_3|^2 > 0$$

or that

$$a_{11}\left|x_1 + x_2\frac{a_{12}}{a_{11}} + x_3\frac{a_{13}}{a_{11}}\right|^2 + \left(a_{22} - \frac{a_{12}\bar{a}_{12}}{a_{11}}\right)|x_2|^2 +$$
$$+ \left(a_{23} - \frac{\bar{a}_{12}a_{13}}{a_{11}}\right)\bar{x}_2 x_3 + \left(\bar{a}_{23} - \frac{a_{12}\bar{a}_{13}}{a_{11}}\right)x_2\bar{x}_3$$
$$+ \left(a_{33} - \frac{a_{13}\bar{a}_{13}}{a_{11}}\right)|x_3|^2 > 0$$

And we see upon taking

$$x_1 + x_2\frac{a_{12}}{a_{11}} + x_3\frac{a_{13}}{a_{11}} = 0$$

that the quadratic form in x_2 and x_3 must be positive definite. It follows, upon applying the result for 2 x 2 matrices obtained before, that

$$\begin{vmatrix} a_{22} - \dfrac{a_{12}\bar{a}_{12}}{a_{11}} & a_{23} - \dfrac{\bar{a}_{12}a_{13}}{a_{11}} \\ \bar{a}_{23} - \dfrac{a_{12}\bar{a}_{13}}{a_{11}} & a_{33} - \dfrac{a_{13}\bar{a}_{13}}{a_{11}} \end{vmatrix} > 0$$

The above condition can be stated better by considering the determinant

$$\begin{vmatrix} a_{11} & a_{12} & a_{13} \\ \bar{a}_{12} & a_{22} & a_{23} \\ \bar{a}_{13} & \bar{a}_{23} & a_{33} \end{vmatrix} = \begin{vmatrix} a_{11} & a_{12} & a_{13} \\ 0 & a_{22} - \dfrac{a_{12}\bar{a}_{12}}{a_{11}} & a_{23} - \dfrac{\bar{a}_{12}a_{13}}{a_{11}} \\ 0 & \bar{a}_{23} - \dfrac{a_{12}\bar{a}_{13}}{a_{11}} & a_{33} - \dfrac{a_{13}\bar{a}_{13}}{a_{11}} \end{vmatrix} > 0$$

Consequently, if we follow an inductive proof, we deduce that all principal minors must be all positive.

To show that A possesses positive eigenvalues, we choose x to be an eigenvector of A, and obtain

$$u^{i*}Au^i = \lambda_i u^{i*}u^i > 0, \quad i = 1, \ldots, n$$

which gives

$$\lambda_i > 0, \quad i = 1, \ldots, n$$

Finally, we put A in the form

$$A = T\begin{bmatrix} \lambda_1 & & & \\ & \lambda_2 & & \\ & & \ddots & \\ & & & \lambda_n \end{bmatrix}T^*$$

and choose a matrix $W = \text{diag}\,(\sqrt{\lambda_1}, \sqrt{\lambda_2}, \ldots, \sqrt{\lambda_n})$. Hence

$$A = TWW^*T^* = QQ^*$$

and the proof is complete.

4.7. The eigenvalue problem for non-semi-simple matrices

Not every matrix can be put in a diagonal form, only semi-simple matrices, since their modal matrix is nonsingular. However if the matrix is non-semi-simple, it can be put in a form close to the diagonal form named *Jordan block*. The method is demonstrated as follows: If λ is a non-semi-simple eigenvalue of A of multiplicity equal to m with corresponding linearly dependent eigenvectors $u^1, u^2, \ldots, u^m$, we form the following equations

$$\begin{aligned} Au^1 &= \lambda u^1 \\ A\tilde{u}^2 &= \lambda \tilde{u}^2 + u^1 \\ A\tilde{u}^3 &= \lambda \tilde{u}^3 + \tilde{u}^2 \\ &\vdots \\ A\tilde{u}^m &= \lambda \tilde{u}^m + \tilde{u}^{m-1} \end{aligned}$$

Suppose that $u^1, \tilde{u}^2, \ldots, \tilde{u}^m$ can be shown to be linearly independent; then the above equations, together with the remaining ones concerning the semi-simple eigenvalues which are

$$\begin{aligned} Au^{m+1} &= \lambda_{m+1} u^{m+1} \\ &\vdots \\ Au^n &= \lambda_n u^n \end{aligned}$$

can all be put in the form

$$T^{-1}AT = J$$

since T is nonsingular (see Exercise 4.1.14). J is called the Jordan block for A, and is given by

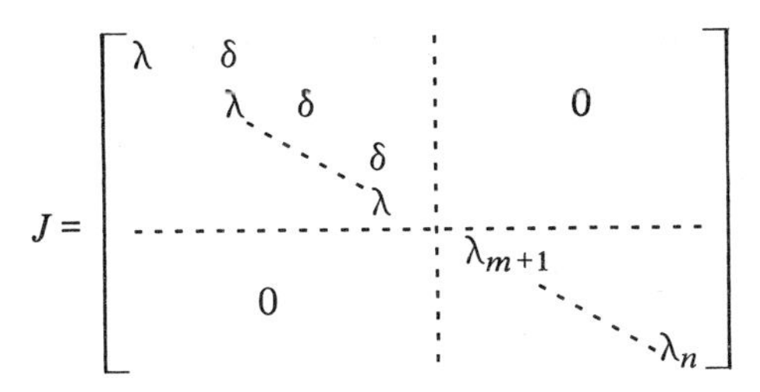

where δ takes the value zero or one. T is the new modal matrix containing the generalized (principal) eigenvectors as follows:

$$T = [u^1 \tilde{u}^2 \ldots \tilde{u}^m u^{m+1} \ldots u^n]$$

To show that $u^1, \tilde{u}^2, \ldots, \tilde{u}^m$ are linearly independent, we rewrite the above equations relating the generalized eigenvectors to the original ones in the following manner

$$(A - \lambda I)\, u^1 = 0$$
$$(A - \lambda I)\, \tilde{u}^2 = u^1$$
$$\vdots$$
$$(A - \lambda I)\, \tilde{u}^m = \tilde{u}^{m-1}$$

The above equations give as well

$$(A - \lambda I)\, u^1 = 0$$
$$(A - \lambda I)^2\, \tilde{u}^2 = (A - \lambda I)\, u^1 = 0$$
$$(A - \lambda I)^3\, \tilde{u}^3 = (A - \lambda I)^2\, \tilde{u}^2 = 0$$
$$\vdots$$
$$(A - \lambda I)^m \tilde{u}^m = 0$$

Now form the linear relation

$$c_1 u^1 + c_2 \tilde{u}^2 + \cdots + c_m \tilde{u}^m = 0$$

and proceed to show that

$$c_1 = c_2 = \cdots = c_m = 0$$

Pre-multiplying the above equation by $(A - \lambda I)^{m-1}$ gives

$$c_m u^1 = 0$$

from which

$$c_m = 0$$

And the process can be repeated to give $c_{m-1} = 0$, and so on until $c_1 = 0$. Thus the proof is complete.

In the above procedure u^1 was chosen out of $u^1, u^2, \ldots, u^m$ and a new set $u^1, \tilde{u}^2, \ldots, \tilde{u}^m$ was constructed. This happens only when $u^1, u^2, \ldots, u^m$ have rank one. If on the other hand, their rank is equal to r $(r < m)$, the generalized eigenvectors will contain r linearly independent vectors of $u^1, u^2, \ldots, u^m$. In this case, the same eigenvalue λ appears in more than one elementary Jordan block, and the matrix A will then be called *derogatory*.

Example. Find a matrix T, such that $T^{-1}AT = J$, with

$$A = \begin{bmatrix} 17 & 0 & -25 \\ 0 & 3 & 0 \\ 9 & 0 & -13 \end{bmatrix}$$

The eigenvalues of A were found by solving the characteristic equation, to give

$$\lambda_1 = 3, \quad \lambda_2 = 2, \quad \lambda_3 = 2$$

The eigenvectors can also be found to be

$$u^1 = \begin{bmatrix} 0 \\ 1 \\ 0 \end{bmatrix}, \quad u^2 = u^3 = \begin{bmatrix} 5 \\ 0 \\ 3 \end{bmatrix}$$

Therefore the generalized eigenvectors will be u^1, u^2 and $\tilde{u}^3$, such that

$$\begin{bmatrix} 15 & 0 & -25 \\ 0 & 1 & 0 \\ 9 & 0 & -15 \end{bmatrix} \tilde{u}^3 = \begin{bmatrix} 5 \\ 0 \\ 3 \end{bmatrix} \Rightarrow \tilde{u}^3 = \begin{bmatrix} 1/3 \\ 0 \\ 0 \end{bmatrix}$$

Hence

$$T = \begin{bmatrix} 0 & 5 & 1/3 \\ 1 & 0 & 0 \\ 0 & 3 & 0 \end{bmatrix}, \quad J = \begin{bmatrix} 3 & 0 & 0 \\ 0 & 2 & 1 \\ 0 & 0 & 2 \end{bmatrix}$$

4.8. The Cayley–Hamilton theorem

The Cayley–Hamilton theorem is of paramount importance in finding functions of matrices. It states that any matrix satisfies its own characteristic polynomial. In other words, if the characteristic polynomial $f(\lambda)$ of A is given by

$$f(\lambda) = |\lambda I - A| = \lambda^n + a_{n-1}\lambda^{n-1} + \cdots + a_0$$

then

$$F(A) = A^n + a_{n-1}A^{n-1} + \cdots + a_0 I = 0_{n,n}$$

Proof. There are many proofs of the Cayley–Hamilton theorem; we give one of the simplest, which of course covers the most general case when A is non-semi-simple. Let J be the Jordan block for A, given by

$$J = \left[\begin{array}{cccc:ccc} \lambda & 1 & & & & & \\ & \lambda & 1 & & & 0 & \\ & & \ddots & 1 & & & \\ & & & \lambda & & & \\ \hdashline & & & & \lambda_{m+1} & & \\ & 0 & & & & \ddots & \\ & & & & & & \lambda_n \end{array}\right]$$

where A is assumed non-derogatory without loss of generality. λ is a non-semi-simple eigenvalue of A with multiplicity equal to m. Hence

$$A = TJT^{-1}$$

also

$$A^2 = TJT^{-1}TJT^{-1} = TJ^2T^{-1},$$

where J^2 is given by

$$J^2 = \left[\begin{array}{cccc:ccc} \lambda^2 & 2\lambda & & & & & \\ & \lambda^2 & 2\lambda & & & 0 & \\ & & \ddots & 2\lambda & & & \\ & & & \lambda^2 & & & \\ \hdashline & & & & \lambda_{m+1}^2 & & \\ & 0 & & & & \ddots & \\ & & & & & & \lambda_n^2 \end{array}\right]$$

In general

$$A^k = TJ^kT^{-1}$$

where J^k is given by

$$J^k = \left[\begin{array}{cccc:ccc} \lambda^k & k\lambda^{k-1} & \cdots & (k_{m-1})\lambda^{k-m+1} & & & \\ & \lambda^k & & \vdots & & 0 & \\ & & \ddots & \lambda^k & & & \\ \hdashline & & & & \lambda_{m+1}^k & & \\ & 0 & & & & \ddots & \\ & & & & & & \lambda_n^k \end{array}\right]$$

Substituting A^k in $F(A)$, we obtain

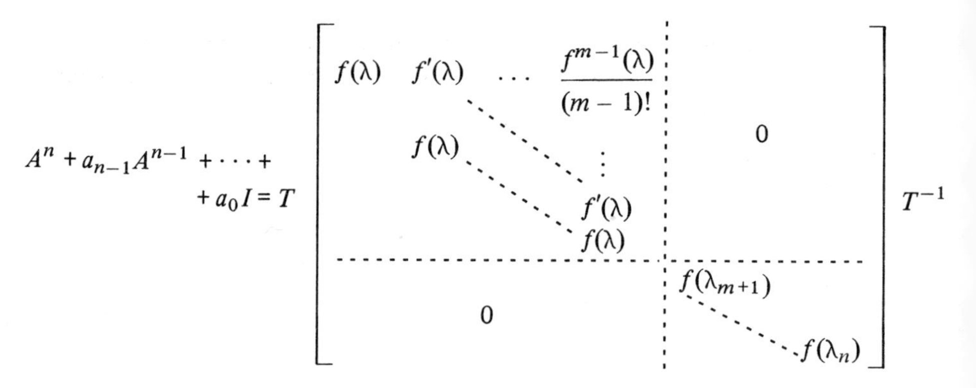

However

$$f(\lambda) = f(\lambda_{m+1}) = \cdots = f(\lambda_n) = 0$$

since λ_i satisfies the characteristic polynomial. Also because λ is of multiplicity m, we also have

$$f'(\lambda) = f''(\lambda) = \cdots = f^{(m-1)}(\lambda) = 0$$

Hence

$$F(A) = 0$$

which completes the proof.

Example. Check the Cayley–Hamilton theorem for A, where

$$A = \begin{bmatrix} 1 & 0 & -2 \\ 1 & 1 & 0 \\ 0 & 1 & 2 \end{bmatrix}$$

The characteristic polynomial is

$$f(\lambda) = \lambda(\lambda^2 - 4\lambda + 5)$$

Therefore

$$F(A) = A(A^2 - 4A + 5I)$$

$$= \begin{bmatrix} 1 & 0 & -2 \\ 1 & 1 & 0 \\ 0 & 1 & 2 \end{bmatrix} \left\{ \begin{bmatrix} 1 & -2 & -6 \\ 2 & 1 & -2 \\ 1 & 3 & 4 \end{bmatrix} - 4 \begin{bmatrix} 1 & 0 & -2 \\ 1 & 1 & 0 \\ 0 & 1 & 2 \end{bmatrix} + \begin{bmatrix} 5 & 0 & 0 \\ 0 & 5 & 0 \\ 0 & 0 & 5 \end{bmatrix} \right\}$$

$$= \begin{bmatrix} 1 & 0 & -2 \\ 1 & 1 & 0 \\ 0 & 1 & 2 \end{bmatrix} \begin{bmatrix} 2 & -2 & 2 \\ -2 & 2 & -2 \\ 1 & -1 & 1 \end{bmatrix} = \begin{bmatrix} 0 & 0 & 0 \\ 0 & 0 & 0 \\ 0 & 0 & 0 \end{bmatrix}$$

4.9. The minimal polynomial of a matrix

Although the Cayley–Hamilton theorem applies to any matrix A, some matrices not only satisfy their own characteristic polynomial but also satisfy a lower-order polynomial (the minimal polynomial of A) which is a divisor of the characteristic polynomial.

To explain this further, let

$$|\lambda I - A| = \lambda^n + a_{n-1}\lambda^{n-1} + \cdots + a_0$$

then

$$A^n + a_{n-1}A^{n-1} + \cdots + a_0 I = \prod_{i=1}^{n} (A - \lambda_i I) = 0$$

by the Cayley–Hamilton theorem. And if r is the least positive number for which

$$\prod_{i=1}^{r} (A - \lambda_i I) = 0, \quad r < n$$

then

$$\prod_{i=1}^{r} (\lambda - \lambda_i)$$

is called the *minimal polynomial* of A.

Example. Obtain the minimal polynomial for A, where

$$A = \begin{bmatrix} 5 & -2 & -4 \\ -2 & 2 & 2 \\ -4 & 2 & 5 \end{bmatrix}$$

The characteristic polynomial is

$$f(\lambda) = (\lambda - 10)(\lambda - 1)(\lambda - 1)$$

Hence

$$(A - 10I)(A - I)(A - I) = 0$$

by the Cayley–Hamilton theorem.
However

$$(A - 10I)(A - I) = \begin{bmatrix} -5 & -2 & -4 \\ -2 & -8 & 2 \\ -4 & 2 & -5 \end{bmatrix} \begin{bmatrix} 4 & -2 & -4 \\ -2 & 1 & 2 \\ -4 & 2 & 4 \end{bmatrix} = \begin{bmatrix} 0 & 0 & 0 \\ 0 & 0 & 0 \\ 0 & 0 & 0 \end{bmatrix}$$

which is also zero. Therefore the minimal polynomial is

$$(\lambda - 10)(\lambda - 1) = \lambda^2 - 11\lambda + 10$$

The reader should be aware of a common mistake, namely to think that the minimal polynomial is composed of all prime divisors of the characteristic polynomial, in other words being equal to

$$\prod_{i=1}^{s} (\lambda - \lambda_i)$$

where s is the number of distinct eigenvalues. But this is not true, for example, of the four matrices:

$$\begin{bmatrix} 2 & & & \\ & 2 & & \\ & & 2 & \\ & & & 2 \end{bmatrix}, \begin{bmatrix} 2 & & & \\ & 2 & & \\ & & 2 & 1 \\ & & & 2 \end{bmatrix}, \begin{bmatrix} 2 & & & \\ & 2 & 1 & \\ & & 2 & 1 \\ & & & 2 \end{bmatrix},$$

$$\begin{bmatrix} 2 & 1 & & \\ & 2 & 1 & \\ & & 2 & 1 \\ & & & 2 \end{bmatrix}$$

which each has the same characteristic polynomial equal to

$$(\lambda - 2)^4$$

However, they differ in their minimal polynomial. For the first matrix, it is $(\lambda - 2)$ since

$$A - 2I = 0,$$

for the second, it is $(\lambda - 2)^2$, since

$$(A - 2I)^2 = 0,$$

for the third, it is $(\lambda - 2)^3$, since

$$(A - 2I)^3 = 0$$

and for the fourth it is $(\lambda - 2)^4$, since only

$$(A - 2I)^4 = 0$$

However, if A is semi-simple, we do obtain a simple result stated in the following theorem:

THEOREM 1. If A is semi-simple, the minimal polynomial of A is equal to $\prod_{i=1}^{s} (\lambda - \lambda_i)$, where s is the number of distinct eigenvalues.

Proof. If the matrix polynomial

$$\prod_{i=1}^{s} (A - \lambda_i I)$$

is zero, the theorem is proved. And we have

$$\begin{aligned}\prod_{i=1}^{s} (A - \lambda_i I) &= T[T^{-1}(A - \lambda_1 I)TT^{-1}(A - \lambda_2 I)T \ldots T^{-1}(A - \lambda_s I)T]\, T^{-1} \\ &= T(D_\lambda - \lambda_1 I)(D_\lambda - \lambda_2 I) \ldots (D_\lambda - \lambda_s I)T^{-1} \\ &= 0\end{aligned}$$

THEOREM 2. If A is non-semi-simple and non-derogatory, the minimal polynomial is equal to the characteristic polynomial.

The proof of theorem 2 is left as an exercise for the reader. Of course if A is non-semi-simple and derogatory, the minimal polynomial is an order lower than the characteristic polynomial but larger in order than just $\prod_{i=1}^{s} (\lambda - \lambda_i)$. To calculate it, we use the traditional method of testing $\prod_{i=1}^{r} (A - \lambda_i I)$. In the next chapter we explain a more efficient method for calculating the minimal polynomial.

4.10. Spectral representation of a matrix

The spectral representation of a matrix enables us to represent any square matrix A in the form

$$A = \sum_{i=1}^{n} \lambda_i E_i$$

It is behind many physical applications, like the concept of normal modes in mechanics and electrical engineering systems. If A is non-semi-simple, the above representation is a little bit changed.

Let A be a general square matrix, preferably non-semi-simple (for considering the general case). Assume, without loss of generality, that the generalized eigenvectors of A are $u^1, u^2, \ldots, u^m, u^{m+1}, \ldots, u^n$; where the first m vectors are the generalized eigenvectors corresponding to a non-semi-simple eigenvalue of multiplicity m, and the rest are eigenvectors corresponding to the semi-simple eigenvalues. Let x be any vector in $\mathbb{C}^n$; then expanding x in terms of the eigenvectors, we obtain

$$x = \sum_{i=1}^{n} c_i u^i$$

$$= \sum_{i=1}^{n} \frac{\langle v^i, x \rangle}{\langle v^i, u^i \rangle} u^i$$

where $v^1, v^2, \ldots, v^n$ are the reciprocal bases for $u^1, u^2, \ldots, u^n$. And if we normalize such that $\langle v^i, u^i \rangle = 1$, we obtain

$$x = \sum_{i=1}^{n} \langle v^i, x \rangle u^i$$

$$= \left(\sum_{i}^{n} u^i \rangle\langle v^i \right) x$$

where the matrix $u^i \rangle\langle v^i = u^i v^{i*} = E_i$ is called the ith *eigenprojection of A* corresponding to the eigenvalue λ_i. Note that E_i is an idempotent matrix since

$$E_i^k = u^i \rangle\langle v^i u^i \rangle\langle v^i \ldots u^i \rangle\langle v^i = u^i \rangle\langle v^i = E_i$$

Also the expansion of x can be written as

$$\left(I - \sum_{i=1}^{n} E_i \right) x = 0, \quad \forall x$$

implying that

$$\sum_{i=1}^{n} E_i = I$$

Another property of the eigenprojection is that

$$E_i E_j = u^i \rangle\langle v^i u^j \rangle\langle v^j = 0, \quad i \neq j$$

Now pre-multiplying the identity

$$I = \sum_{i=1}^{n} E_i = \sum_{i=1}^{n} u^i \rangle\langle v^i$$

by the matrix A, we obtain

$$A = \sum_{i=1}^{n} \lambda_i E_i + \sum_{i=1}^{m-1} u^i \rangle\langle v^{i+1}$$

where the second matrix on the right-hand side is a nilpotent matrix since

$$\left(\sum_{i=1}^{m-1} u^i \rangle\langle v^{i+1} \right)^m = 0.$$

Note that this term does not exist if A is semi-simple.

The spectral representation of a matrix shows that any matrix A can additively decompose into two matrices, one idempotent and the other nilpotent. In other words A can be written as the addition of two matrices, one diagonalizable (which one is this?), and the other non-diagonalizable. The reader may observe that the spectral representation of a matrix is another way of writing

$$A = \begin{bmatrix} u^1 u^2 \ldots u^n \end{bmatrix} \left[\begin{array}{ccc|ccc} \lambda & \delta & & & & \\ & \lambda & \delta & & & \\ & & \ddots & \delta & & \\ & & & \lambda & & \\ \hline & & & & \lambda_{m+1} & \\ & & & & & \ddots \\ & & & & & \lambda_n \end{array}\right] \begin{bmatrix} v^{1*} \\ v^{2*} \\ \vdots \\ v^{n*} \end{bmatrix}$$

where δ is either zero or one.

Sometimes, the eigenprojections corresponding to a multiple eigenvalue λ, whether semi-simple or not, are added together, and the spectral expansion of A takes the form

$$A = \sum_{i=1}^{s} \lambda_i E_i + \sum_{i=1}^{m-1} u^i \rangle\langle v^{i+1}$$

where s is the number of distinct eigenvalues.

The spectral representation (often called *dyadic expansion*) of a matrix appears in many kinds of application. Apart from its main application of computing different matrix functions (see Section 5.6), it also has extensive numerical use. This often appears in connection with the spectral radius of A. For example to compute A^k for large k, we can use the formula

$$A^* \approx \lambda_1^k E_1$$

where λ_1 is the eigenvalue of maximum modulus, i.e.

$$|\lambda_1| = s(A)$$

and is assumed semi-simple for simplicity. E_1 is its corresponding eigenprojection. This formula could be used as an alternative to the power method if only λ_1 is required. λ_1 is obtained from the formula

$$\operatorname{tr} A^k \approx \lambda_1^k \operatorname{tr} E_1 = \lambda_1^k \operatorname{tr} u^1 \rangle\langle v^1 = \lambda_1^k \operatorname{tr}\langle v^1, u^1 \rangle = \lambda_1^k$$

giving

$$\lambda_1 \approx \sqrt[k]{\operatorname{tr} A^k}$$

if k is taken large enough.

If A is Hermitian or real symmetric, the spectral representation takes the simple form

$$A = \sum_{i=1}^{s} \lambda_i E_i = \sum_{i=1}^{s} \lambda_i u^i \rangle \langle u^i$$

calling the attention of the reader to its similarity with the deflation techniques of the power method (see Section 4.3). For after calculating λ_1 and u^1, we construct the matrix

$$A - \lambda_1 u^1 u^{1*} = \sum_{i=2}^{s} \lambda_i u^i \rangle \langle u^i$$

whose rank is $n - 1$ if A is nonsingular. This matrix has the same eigenvalues of A except that λ_1 is transferred to the origin, and the same eigenvectors of A. Therefore the power method can be reapplied to this matrix to obtain λ_2 and u^2 and so forth. In each step, we actually exhaust the matrix A by reducing its rank.

Another useful numerical application reported by Wilkinson (1965), p. 171, is when an approximate eigenvalue λ and corresponding approximate eigenvector x have been found, and we may wish to bound the error in terms of the residual vector

$$z = Ax - \lambda x$$

Substituting for A with its spectral representation and assuming for simplicity that A is semi-simple, we obtain

$$\begin{aligned} z &= \left(\sum_{i=1}^{n} \lambda_i E_i \right) x - \lambda x \\ &= \left(\sum_{i=1}^{n} \lambda_i E_i \right) x - \lambda \left(\sum_{i=1}^{n} E_i \right) x \\ &= \left(\sum_{i=1}^{n} (\lambda_i - \lambda) E_i \right) x \end{aligned}$$

This equation can be reduced to the simple form

$$z = T \begin{bmatrix} \lambda_1 - \lambda & & & \\ & \lambda_2 - \lambda & & \\ & & \ddots & \\ & & & \lambda_n - \lambda \end{bmatrix} T^{-1} x = T(D_\lambda - \lambda I) T^{-1} x$$

where T is the modal matrix of A. Rewriting the above equation in the form

$$x = T(D_\lambda - \lambda I)^{-1} T^{-1} z$$

and on taking the norm of both sides we obtain

$$\| x \|_2 \leqslant \| T \|_2 \| T^{-1} \|_2 \| (D_\lambda - \lambda I)^{-1} \|_2 \| z \|_2$$
$$= \frac{\| T \|_2 \| T^{-1} \|_2 \| z \|_2}{\min_i | \lambda_i - \lambda |}$$

i.e.

$$\min_i | \lambda - \lambda_i | \leqslant \mathscr{K} \frac{\| z \|_2}{\| x \|_2}$$

where

$$\mathscr{K} = \| T^{-1} \|_2 \| T \|_2 \geqslant 1$$

is called the *spectral condition number* of A. When A is Hermitian, T is unitary and $\mathscr{K} = 1$, meaning that Hermitian and real symmetric matrices have well-conditioned eigenproblems in this sense. If A is not Hermitian, it means that λ and the corresponding eigenvector x are not usually calculated accurately. One very powerful algorithm for calculating accurately the eigenvectors is the one devised by Wielandt called the inverse iteration method. Assume that a good approximation λ is already known for one of the eigenvalues $\lambda_1, \ldots, \lambda_n$ of A, say λ_j. Let T be the modal matrix of A; then

$$(A - \lambda I)^{-1} = T(D_\lambda - \lambda I)^{-1} T^{-1}$$
$$= \sum_{i=1}^{n} \frac{E_i}{\lambda_i - \lambda}$$

Now choose a vector y and expand it in terms of the eigenvectors of A, i.e.

$$y = \sum_{i=1}^{n} c_i u^i$$

Post-multiplying $(A - \lambda I)^{-1}$ by y, we obtain

$$y^1 = (A - \lambda I)^{-1} y = \sum_{i=1}^{n} \frac{c_i u^i}{\lambda_i - \lambda}$$

Repeating the iterations, we obtain

$$y^2 = (A - \lambda I)^{-1} y^1 = \sum_{i=1}^{n} \frac{c_i u^i}{(\lambda_i - \lambda)^2}$$
$$\vdots$$
$$y^m = (A - \lambda I)^{-1} y^{m-1} = \sum_{i=1}^{n} \frac{c_i u^i}{(\lambda_i - \lambda)^m}$$

If m is chosen large enough, we find that y^m becomes, by iteration, $\dfrac{c_j u^j}{(\lambda_j - \lambda)^m}$, since

$$y^m = \frac{1}{(\lambda_j - \lambda)^m}\left(c_j u^j + \sum_{i \neq j} \frac{(\lambda_j - \lambda)^m}{(\lambda_i - \lambda)^m} c_i u^i\right)$$

The factors $|(\lambda_j - \lambda)/(\lambda_i - \lambda)|$ are usually small, even if the eigenvalues are not well separated. Therefore convergence is rapid, and u^j is calculated accurately.

The inverse iteration method is used when one tries to find a numerically acceptable eigenvector u^j better than x, if a numerically acceptable eigenvalue λ is already known. Such a method is by far the most powerful and accurate method for computing eigenvectors; see Wilkinson (1965), p. 622.

Exercises 4.10

1. Find a matrix T, such that $T^{-1}AT = J$, if

$$A = \begin{bmatrix} 0 & 1 & 0 \\ -4 & 4 & 0 \\ 0 & 0 & 2 \end{bmatrix}, \quad A = \begin{bmatrix} 1 & 1 & -1 \\ 0 & 2 & -2 \\ 0 & 0 & 1 \end{bmatrix}$$

2. Find a matrix T, such that $T^*AT = D$, if

$$A = \begin{bmatrix} 0 & -2 & 0 \\ 2 & -i & 0 \\ 0 & 0 & i \end{bmatrix}, \quad A = \begin{bmatrix} 0 & 2 & -1 \\ 2 & 5 & -6 \\ -1 & -6 & 8 \end{bmatrix}$$

3. Show that the eigenvalues of a real skew-symmetric or skew-Hermitian matrix are zero or pure imaginary, and two eigenvectors corresponding to two distinct eigenvalues are orthogonal.
4. A matrix A is called *symmetrizable* if $LA = A^T L$, for some positive definite matrix L. Show that A has real eigenvalues and two eigenvectors corresponding to two distinct eigenvalues are L orthogonal, i.e. $\langle u^i, Lu^j \rangle = 0, i \neq j$.
5. A matrix A is called *skew-symmetrizable* if $LA = -A^T L$, for some positive definite matrix L. Show that A has pure imaginary eigenvalues, and two eigenvectors corresponding to two distinct eigenvalues are L orthogonal.
6. If P is positive definite, A negative semidefinite and B skew-Hermitian, show that the eigenvalues of $P^{-1}(A + B)$ are located in the left-hand side of the λ-plane. $P^{-1}(A + B)$ may represent the state matrix of any $R^+ - L^+ - C^+$ network.
7. If K is positive definite and L negative semidefinite, show that $K^{-1}L$ has real non-positive eigenvalues. $K^{-1}L$ may represent the state matrix of any $R^+ - L^+$ or $R^+ - C^+$ network.
8. If F is positive definite and J is skew-Hermitian, show that the eigenvalues of $F^{-1}J$ are pure imaginary. $F^{-1}J$ may represent the state matrix of any $L^+ - C^+$ network.
9. If A is real symmetric positive definite, compute the volume of the ellipsoid $\langle x, Ax \rangle = 1$.
10. If A and B are real symmetric, with B positive definite, show that the roots of $|A - \lambda B| = 0$ are real.
11. If A and B are both real symmetric, show that the eigenvalues of AB are real, provided that at least one of them is either positive or negative definite.
12. If A, B and C are all positive definite, show that the roots of $|A\lambda^2 + B\lambda + C| = 0$

have negative real parts. A, B and C may represent the mass matrix, damping matrix and stiffness matrix respectively of a non-conservative mechanical system.

13. Let A be a real symmetric matrix and let $u^1, u^2, \ldots, u^n$ be an orthonormal set of characteristic vectors of A corresponding to the characteristic roots $\lambda_1, \lambda_2, \ldots, \lambda_n$. Show that the solution of $(A - \lambda I)x = b$ is

$$x = \sum_{j=1}^{n} \frac{\langle u^j, b \rangle}{\lambda_j - \lambda} u^j, \quad \lambda_j \neq \lambda.$$

Show how to obtain x if A is not symmetric.

14. The set of vectors $u^1, u^2, \ldots, u^n$ is called a set of *conjugate directions* with respect to a positive matrix A, if $\langle u^i, Au^j \rangle = 0, i \neq j$.

(a) Show that $u^1, \ldots, u^n$ are linearly independent.

(b) If $x = \sum_{i=1}^{n} c_i u^i$, show how to obtain c_i.

(c) If $\langle u^i, Au^i \rangle = 1$, show that $\langle x, Ax \rangle = \sum_{i=1}^{n} |c_i|^2$.

(d) Show that the eigenvectors of A form a set of conjugate directions for A.

15. The partial differential equation $au_{xx} + bu_{xy} + cu_{yy} = f(x, y)$ is called elliptic, parabolic or hyperbolic, depending on its characteristic equation $a\lambda^2 + b\lambda s + cs^2 = F$. Find conditions for a, b and c in each case. *Hint*: consider eigenvalues of

$$\begin{bmatrix} a & b/2 \\ b/2 & c \end{bmatrix}$$

16. If A is real symmetric, find the condition on ϵ such that $I + \epsilon A$ is positive definite.

17. If A is real symmetric positive definite, and B real skew-symmetric, show that $A + B$ is nonsingular.

18. Show that A^*A is positive definite if A is nonsingular.

19. Show that if A is of order (m, n), $m > n$, $\rho(A) = n$, then A^*A is nonsingular positive definite. Is this still true for $m < n$ even if $\rho(A) = m$?

20. If $\begin{bmatrix} A \\ \hdashline B \end{bmatrix}$ is nonsingular with $AB^* = 0$, show that $\begin{bmatrix} AP \\ \hdashline BQ \end{bmatrix}$ is also nonsingular if $P^{-1}Q$ is either positive or negative definite. For a proof and further corollaries see Deif (1980). This theorem has various applications in engineering systems. Relevant applications exist in truss problems and electrical networks, where P and Q may represent the stiffness matrix or resistive matrix, which are generally positive definite. For example for a general resistive network, the two Kirchhoff's Laws can be written in the form

$$\alpha i = I$$

$$\beta v = E$$

where i and v are respectively the branches' currents and voltages and of dimensions equal to the number of branches b; α and β are respectively the cut-set matrix and tie-set matrix and of dimensions $(n - 1, b)$ and (l, b), n and l being the number of nodes and loops and satisfying $l + n - 1 = b$. I and E are the current and voltage sources. From electrical network theory, $\alpha\beta^T = 0$ and the matrix $\left[\begin{array}{c}\alpha \\ \hline \beta\end{array}\right]$ is nonsingular if both nodes and loops are chosen independently. Therefore, to analyse the network, we substitute $v = Ri$, where R is a diagonal matrix of positive elements, representing the values of the branch resistors, i.e. R is positive definite. The final equations become

$$\left[\begin{array}{c}\alpha \\ \hline \beta R\end{array}\right] i = \left[\begin{array}{c}I \\ \hline E\end{array}\right]$$

where the matrix on the left-hand side is nonsingular according to the above theorem. To apply the theorem to an example, refer to Exercise 3.10.7.

21. If M and K are two Hermitian matrices and M is positive definite, show that there exists a matrix C, such that $C^*MC = I$, $C^*KC = \text{diag}(\lambda_1 \lambda_2, \ldots, \lambda_n)$, where λ_i is real and satisfying the eigenvalue problem $Kc^i = \lambda_i Mc^i$, in which c^i is a column vector of C. This method is called *simultaneous diagonalization.*

22. If A is Hermitian, its *signature* is the difference between the number of positive and negative eigenvalues. Show that the signature is invariant under a congruent transformation, i.e. A and K^*AK have the same signature for any nonsingular matrix K.

23. If P is positive definite, we define the norm of x as $\| x \| = \sqrt{\langle x, Px \rangle}$; this is called the *Riemannian metric*, used in differential geometry, and the Euclidian norm is a special case when $P = I$. Show that the Riemannian metric satisfies the conditions of a vector norm.

24. If A is positive definite, show that:
(a) A is nonsingular.
(b) A^k (k integer) is positive definite.
(c) $-A$ is negative definite.
(d) M^*AM is positive definite for any nonsingular M.

25. If A is negative definite, show that:
(a) Eigenvalues of A are all negative.
(b) A^k is positive definite for k even and negative definite for k odd.
(c) The leading principal minors are alternating in sign i.e.

$$D_1 < 0, \quad D_2 > 0, \quad D_3 < 0 \text{ etc.}$$

26. Show that the quadratic form $\langle x, Ax \rangle$ can be represented in the form

$\langle x, Ax \rangle = \sum_{k=1}^{n} (D_k/D_{k-1})\, y_k^2$, provided that no $D_k = 0$, where $D_0 = 1$ and

$y_k = x_k + \sum_{j=k+1}^{n} c_{kj} x_j$, $k = 1, \ldots, n-1$, $y_n = x_n$ and c_{ij} are rational functions of a_{ij}. Show how this representation produces easily the determinantal criteria for positive or negative definiteness of A. What happens if one or more of the D_k are zero?

27. If A is positive semi-definite, show that

(a) $\langle x, Ax \rangle = \sum_{i=1}^{r} \lambda_i |y_i|^2$, where r is the rank of A.

(b) $A = Q^*DQ$, where Q is a nonsingular matrix and $D = \begin{bmatrix} I_r & 0 \\ 0 & 0 \end{bmatrix}$.

Can you obtain determinantal criteria?

28. If A is positive definite, show that $|A| < a_{11}a_{22} \ldots a_{nn}$.

29. If A and B are both positive definite and C and D are both negative definite and all matrices commute, show that

(a) AB is positive definite.

(b) CD is positive definite.

(c) AC is negative negative.

30. Show that, for a normal matrix, two eigenvectors corresponding to two distinct eigenvalues are orthogonal.

31. If A is normal, show that $\lambda(AA^*) = |\lambda(A)|^2$ and $\lambda(A + A^*) = \lambda(A) + \bar{\lambda}(A)$.

32. If A is normal, show that $A = B + iC$ $(i = \sqrt{-1})$, where B and C commute. Show also that $\lambda(A) = \lambda(B) + i\lambda(C)$.

33. If A is normal and has real eigenvalues, show that A is Hermitian.

34. Show that every nonsingular matrix A can be written as $A = HU$, where H is Hermitian positive definite and U is unitary. *Hint*: $H^2 = AA^*$, $U = H^{-1}A$. Show also that if A is normal then H and U commute.

35. Show that a matrix A is diagonalizable if and only if there exists a positive definite Hermitian matrix P such that PAP^{-1} is a normal matrix.

36. If A and B are normal, show that AB is normal if and only if each commutes with the H matrix of the other. By the H matrix of A we mean the positive non-negative square root of AA^*.

37. Introduce the notion of positive definiteness for normal matrices as follows: a normal matrix A is called positive definite if Re $x^*Ax > 0, \forall x \neq 0$. Show that a necessary and sufficient condition for A to be positive definite is that any one of the following conditions is satisfied:

(a) Re $\lambda > 0$.

(b) All leading principal minors of $A + A^*$ are positive.

(c) $A = QDQ^*$, where Q is a nonsingular matrix and $D = \text{diag}(e^{i\theta_1}, \ldots, e^{i\theta_n})$ with $-\pi/2 < \theta_i < \pi/2$; where $e^{i\theta_1}, \ldots, e^{i\theta_n}$ are the eigenvalues of a unitary matrix U such that $A = HU = UH$, with H Hermitian and positive definite; and where Q^*Q is diagonal.

38. Let A be an arbitrary complex matrix of order (m, n); show that there exists a unitary matrix U of order (m, m) and a unitary matrix V of order (n, n) such that $U^*AV = Z$ is an $m \times n$ diagonal matrix of the form $Z = \begin{bmatrix} D & 0 \\ 0 & 0 \end{bmatrix}$,

$D = \text{diag}(\sigma_1, \ldots, \sigma_r)$, $\sigma_1, \ldots, \sigma_r > 0$, where $\sigma_1, \ldots, \sigma_r$ are the non-vanishing singular values of A and r is the rank of A. *Hint*: the columns of U are m orthonormal eigenvectors of AA^*, while those of V are n orthonormal eigenvectors of A^*A. Use also the theorem to obtain R and P matrices such that

$$RAP = \begin{bmatrix} I_r & 0 \\ 0 & 0 \end{bmatrix} = \text{normal form of } A.$$

39. If H_1 and H_2 are both Hermitian and at least one is positive definite, show that H_1H_2 is semi-simple.

40. If U is unitary, show that $|\lambda(U)| = 1$.

41. If $AB = BA$, show that $\lambda(A + B) = \lambda(A) + \lambda(B)$, $\lambda(AB) = \lambda(A)\lambda(B)$.

42. Obtain the spectral representations for the matrices

$$A = \begin{bmatrix} 1 & \sqrt{8} & 0 \\ \sqrt{8} & -1 & 0 \\ 0 & 0 & 3 \end{bmatrix}, \quad A = \begin{bmatrix} \cos\theta & \sin\theta \\ -\sin\theta & \cos\theta \end{bmatrix}, \quad \theta = 0, \frac{\pi}{2}$$

as well as their minimal polynomials.

43. Show that A and $L^{-1}AL$ have same minimal polynomial.

44. If E_i is an eigenprojection of A, show that $\rho(E_i)$ = multiplicity of the eigenvalue λ_i.

45. Show that A commutes with its nilpotent part.

46. Show that the spectral representation of a matrix is unique.

47. Show that any matrix B can be expanded into the form $B = \sum_{i,j} a_{ij}E_{ij}$, with $E_{ij} = u^i\rangle\langle v^j$, where u^i and v^j are eigenvectors and reciprocals of a matrix A. Show how to obtain a_{ij}.

48. Show that the eigenprojection E_i of A can be easily obtained from the formula $E_i = \prod_{k \neq i} \dfrac{A - \lambda_k I}{\lambda_i - \lambda_k}$, if A is semi-simple.

49. *The method of Krylov* for calculating the characteristic polynomial of a matrix proceeds as follows: choose any vector x^0 and expand x^0 in terms of the eigenvectors of A; i.e. $x^0 = \sum_{i=1}^{n} c_i u^i$. Then by pre-multiplying by A, A^2 and so forth, we obtain $x^1 = Ax^0 = \sum_{i=1}^{n} c_i\lambda_i u^i$, $x^2 = Ax^1 = \sum_{i=1}^{n} c_i\lambda_i^2 u^i, \ldots, x^n = Ax^{n-1} = \sum_{i=1}^{n} c_i\lambda_i^n u^i$. Then we multiply the equations successively by $a_0, a_1, \ldots, a_n$, which are the coefficients of the characteristic polynomial, to obtain

$$a_0x^0 + a_1x^1 + \cdots + a_nx^n = \sum_{i=1}^{n} c_iu^i(a_0 + a_1\lambda_i + \cdots + a_n\lambda_i^n) = 0$$

This gives n linear simultaneous equations to solve for $a_0, a_1, \ldots, a_n$. In the above derivation the matrix A was in fact assumed to be semi-simple. Does the same result hold if A is non-semi-simple? Note also that if a solution is obtained such that $a_{m+1}, \ldots, a_n = 0$, $a_0 + a_1\lambda + \cdots + a_m\lambda^m$ is the minimal polynomial. The above equations can also lead to an easy proof of the Cayley–Hamilton theorem by substituting for $x^1 = Ax^0, x^2 = Ax^1 = A^2x^0, \ldots$ such that

$$a_0x^0 + a_1x^1 + \cdots + x_nx^n = (a_0I + a_1A + \cdots + a_nA^n)\,x^0 = 0, \forall x^0.$$

The reader is asked to make a comparison between this method and the one of Leverrier. Show also how to obtain the eigenvectors using Krylov's method. Hence use the method to obtain the minimal polynomial of

$$A = \begin{bmatrix} 5 & -2 & -4 \\ -2 & 2 & 2 \\ -4 & 2 & 5 \end{bmatrix}$$

For more information about Krylov's method, refer to Faddeeva (1959), p. 149.
50. Prove *Kato's lemma*: if A is Hermitian and $(A - aI)(A - bI)$ is positive definite, A has no eigenvalues in the interval $[a, b]$.
51. If A is any square complex matrix, show that

(a) $\sum_i |\lambda_i|^2 \leqslant \sum_{i,j} |a_{ij}|^2$

(b) $\sum_i |\mathrm{Re}\,\lambda_i|^2 \leqslant \sum_{i,j} |(a_{ij} + \bar{a}_{ji})/2|^2$

(c) $\sum_i |\mathrm{Im}\,\lambda_i|^2 \leqslant \sum_{i,j} |(a_{ij} - \bar{a}_{ji})/2|^2$

Hint: use the Schur theorem in Exercise 4.1.25. For which matrix do the equalities hold?
52. If A is any square complex matrix (of order n), show that

(a) $|\lambda| \leqslant n.\ \max_{i,j} |a_{ij}|$

(b) $|\mathrm{Re}\,\lambda| \leqslant n.\ \max_{i,j} |(a_{ij} + \bar{a}_{ji})/2|$

(c) $|\mathrm{Im}\,\lambda| \leqslant n.\ \max_{i,j} |(a_{ij} - \bar{a}_{ji})/2|$

53. Show that $\| A \|_2 = \max_x \dfrac{\| Ax \|_2}{\| x \|_2} = \sqrt{s(A^*A)}$.
54. If A is Hermitian, show that $\| A \|_2 = s(A)$.
55. Show that $\gamma_2(A) = \dfrac{\sigma_1(A)}{\sigma_n(A)}$, where $\sigma_1(A)$ and $\sigma_n(A)$ are, respectively, the largest and smallest singular value of A. Moreover, if A is Hermitian show that $\gamma_2(A) = \dfrac{\max |\lambda(A)|}{\min |\lambda(A)|}$.
56. Show that $\| A \|_2 \leqslant \| A \|_E \leqslant \sqrt{n}\, \| A \|_2$. *Hint*: consider $\| A \|_E^2 = \sum_{i,j} |a_{ij}|^2 = \mathrm{tr}\,(A^*A)$. Draw also a relation between $\| A \|_E$ and $\| A^{-1} \|_2$.
57. If A has n orthogonal eigenvectors, define the Rayleigh quotient for A as $\dfrac{\langle x, Ax \rangle}{\langle x, x \rangle}$. Show that $\lambda_{\max}(A) = \max_x \dfrac{\langle x, Ax \rangle}{\langle x, x \rangle}$. *Hint*: expand x in terms of the eigenvectors of A.
58. If A is any square complex matrix, such that $A = H_1 + iH_2$ with H_1 and H_2 Hermitian, show that

$$\lambda_{\min}(H_1) \leqslant \mathrm{Re}\,\lambda(A) \leqslant \lambda_{\max}(H_1)$$
$$\lambda_{\min}(H_2) \leqslant \mathrm{Im}\,\lambda(A) \leqslant \lambda_{\max}(H_2)$$

Hint: write, for example,

$$\lambda_{\max}(H_1) = \max_x \frac{\langle x, H_1 x \rangle}{\langle x, x \rangle} = \max_x \mathrm{Re} \frac{\langle x, Ax \rangle}{\langle x, x \rangle} \geqslant \mathrm{Re}\,\lambda(A).$$

59. If B is positive semi-definite, show that

$$\lambda(A+B) \geqslant \lambda(A)$$

When does strict inequality hold?
60. In a *Markov matrix* A, the component $a_{ij} > 0$ represents the probability of transition from state i to state j, thus $\sum_j a_{ij} = 1$, for all i. Show that A has the eigenvalue $\lambda_1 = 1$ with multiplicity one, and $|\lambda_i| < 1$ for all other eigenvalues. Deduce the same condition for A^T. Prove that there is a row vector $u^T > 0$ with $\sum_i u_i = 1$, for which $u^T A = u^T$. *Hint*: use $\| A \|_\infty$ to show that $|\lambda_i| \leqslant 1$ and the power method to show that $\lambda_1 = 1$, making a suitable choice for x. Markov matrices are a special case of a broader class called positive matrices important especially in economics; see Bellman (1970), p. 286 and Pease (1965), p. 379.

4.11. Application to integral equations

Whereas differential equations are equations relating the variable to its derivatives, integral equations relate the variable to its integral. An example of an integral equation is the equation

$$x(t) = a + \int_0^t k(\xi) x(\xi) \, d\xi$$

where k is called the *kernel* of the equation, and the problem is to determine $x(t)$. The reader may ask why we do not transform the above integral equation into a differential equation, in which there is much experience in solution techniques. For example the above equation can be transformed into the differential equation

$$\frac{dx}{dt} = k(t)\, x, \quad x(0) = a$$

The reason is that some integral equations cannot be easily transformed into a differential equation. Moreover, some problems can be formulated only by an integral equation. For example, if an external action $f(x)$ is applied to a linear system whose influence function is $h(x, \xi), a < x, \xi < b$, the result of the action which is called output or response is described by the function

$$\Phi(x) = \int_a^b h(x, \xi) f(\xi) \, d\xi$$

so, if Φ and h are given, the problem becomes that of determining $f(x)$ which gives the response $\Phi(x)$. Relevant applications exist in electrical engineering where it is required to determine the voltage source applied to a circuit, so that it gives a certain prescribed current in any branch; h in this case represents the impulse response of the circuit. Another application exists in structures where for instance it is required to find the density of a distributed load along a bar so that it gives a certain corresponding deflection of the bar. The reader should also note that a differential equation is a relationship between the values of given unknown functions on an infinitesimal interval, whereas an integral equation involves a finite interval. Therefore, it appears natural that the investigation of boundary-value problems is connected with integral equations rather than with differential equations.

As an example of an integral equation [Petrovsky (1971), p. 9], consider an elastic thread of length l, which requires a force $c\Delta l$ to increase its length by Δl. Here c is some constant given by Hooke's law. Let the ends of the thread be attached at two fixed points a and b lying on the x axis, such that it is only acted upon by the horizontal tensile force T_0.

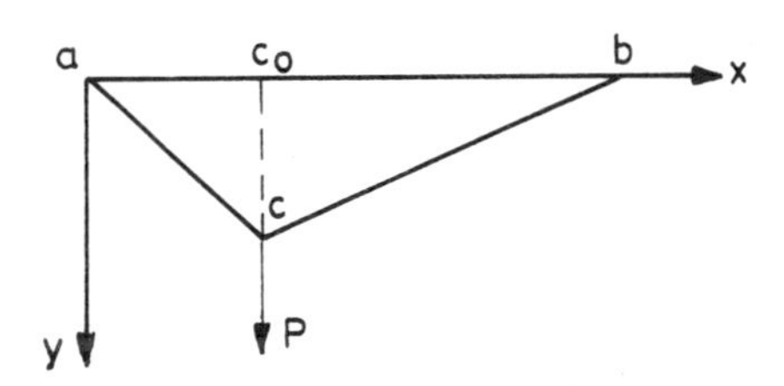

Now suppose that a vertical force P is applied to the thread at point c, for which $x = \xi$. The thread will then assume the shape of the polygonal line acb as in the above figure. We consider $cc_0 = \delta$ to be small compared to ac_0 and bc_0, since $P < T_0$. Now disregard the value δ w.r.t. l; we can take it that the tension of the thread has remained equal to T_0 even under the action of the force P. Projecting the forces vertically we obtain

$$T_0 \frac{\delta}{\xi} + T_0 \frac{\delta}{l - \xi} = P$$

Hence

$$\delta = \frac{P(l - \xi)\xi}{T_0 l}$$

Denoting by $y(x)$ the deflection of the thread at a point with abscissa x, we have

$$y(x) = P.G(x, \xi)$$

where

$$G(x, \xi) = \frac{x(l - \xi)}{T_0 l} \text{ for the segment } ac \, (0 \leqslant x \leqslant \xi)$$

and

$$G(x, \xi) = \frac{(l - x)\xi}{T_0 l} \text{ for the segment } cb \, (\xi \leqslant x \leqslant l)$$

Using these formulae, we can readily verify that

$$G(x, \xi) = G(\xi, x)$$

Suppose that the thread is acted upon by a continuously distributed force with linear density $P(\xi)$; so that a force approximately equal to $P(\xi)\Delta(\xi)$ is acting on the segment between the points $x = \xi$ and $x = \xi + \Delta\xi$. Since the displacements due to the elementary forces $P(\xi)\Delta\xi$ are additive (by the superposition principle), the thread under the action of this force will assume the shape given by

$$y(x) = \int_0^l G(x, \xi)P(\xi)\, d\xi$$

So if $y(x)$ is prescribed to take a given shape $y = y(x)$, we arrive at an integral equation which we need to solve for $P(\xi)$. To complicate the problem a little further, suppose the thread is acted upon by a time-varying force with density at the point ξ equal to $P(\xi) \sin \omega t$.

This will set the thread in motion, and we will assume that in the motion the abscissa of every point of the thread remains unchanged and that the thread executes periodic oscillations described by

$$y = y(x) \sin \omega t$$

where t is the time and $\omega > 0$. Denoting the linear density of mass of the thread at the point ξ by $\rho(\xi)$, we find that at time t the segment of the thread between points ξ and $\xi + \Delta\xi$ is acted upon by the inertial force

$$-\rho(\xi)\Delta\xi \frac{d^2 y}{dt^2} = \rho(\xi)y(\xi)\omega^2 \sin \omega t\, \Delta\xi$$

in addition to the force $P(\xi) \sin \omega t\, \Delta\xi$. Therefore the above integral equation takes the form

$$y(x) \sin \omega t = \int_0^l G(x, \xi)\, [P(\xi) \sin \omega t + \omega^2 \rho(\xi)y(\xi) \sin \omega t]\, d\xi$$

Cancelling $\sin \omega t$ and setting

$$\int_0^l G(x, \xi)P(\xi)\, d\xi = f(x), \quad G(x, \xi)\rho(\xi) = k(x, \xi), \quad \omega^2 = \lambda$$

we get

$$y(x) = f(x) + \lambda \int_0^l k(x, \xi)y(\xi)\, d\xi$$

which is recognized as a Fredholm integral equation of the second kind, and the problem becomes that of solving the equation for $y(x)$.

The basic types of integral equation are the Fredholm type, like

$$\alpha(t)x(t) = f(t) + \lambda \int_a^b k(\xi, t)x(\xi)\, \mathrm{d}\xi$$

and the Volterra type like

$$\alpha(t)x(t) = f(t) + \lambda \int_a^t k(\xi, t)x(\xi)\, \mathrm{d}\xi$$

The reader will notice that by putting

$$k(\xi, t) = 0, \quad \xi > t$$

the Volterra equation becomes a special type of the Fredholm equation. In this section we shall discuss only the Fredholm integral equations. These equations are divided into classes. If $\alpha = 0$, the equations are called Fredholm equations of the first kind, whereas if $\alpha = 1$, they are of the second kind; for other values of α, the equations are of the third kind. However by suitably writing an equation of the third kind, it can be transformed into the second type. Sometimes we have a symmetric kernel, i.e.

$$k(\xi, t) = k(t, \xi)$$

or Hermitian, i.e.

$$k(\xi, t) = \bar{k}(t, \xi)$$

Such integral equations with real symmetric or Hermitian kernels possess interesting properties exactly as in the theory of matrices. We will postpone the discussion of these special kinds until we present methods of solving the ordinary type, where $k(\xi, t)$ is only a continuous function. Writing again the Fredholm's integral equation of the second kind:

$$x(t) = f(t) + \lambda \int_a^b k(\xi, t)x(\xi)\, \mathrm{d}\xi$$

we find that in almost every case $k(\xi, t)$ can be written in the form

$$k(\xi, t) = \sum_{i=1}^{n} r_i(\xi)q_i(t)$$

Such kernels are called *separable* and the integral equation is termed an integral equation with a degenerate kernel. And we shall give some examples to show how to treat equations with a non-separable kernel. Therefore we begin by solving the integral equation with degenerate kernel – of the second kind since it involves the general approach. Substituting for $k(\xi, t)$ in the integral equation, we obtain

$$x(t) = f(t) + \lambda \int_a^b \sum_{i=1}^n r_i(\xi)q_i(t)x(\xi)\,\mathrm{d}\xi$$

$$= f(t) + \lambda \sum_{i=1}^n q_i(t) \int_a^b r_i(\xi)x(\xi)\,\mathrm{d}\xi$$

Calling

$$\int_a^b r_i(\xi)x(\xi)\,\mathrm{d}\xi = x_i = \text{constant}$$

we obtain

$$x(t) = f(t) + \lambda \sum_{j=1}^n x_j q_j(t)$$

Substituting the last equation in the one before, we get

$$\int_a^b r_i(\xi)\left[f(\xi) + \lambda \sum_{j=1}^n x_j q_j(\xi) \right] \mathrm{d}\xi = x_i, \quad i = 1, \ldots, n$$

from which we obtain, in matrix form, the equation

$$(I - \lambda A)x = b$$

where

$$x^T = [x_1, x_2, \ldots, x_n], \quad a_{ij} = \int_a^b r_i(\xi)q_j(\xi)\,\mathrm{d}\xi, \quad b_i = \int_a^b r_i(\xi)f(\xi)\,\mathrm{d}\xi$$

The reader will realize that the above equations have a unique solution for x, only if

$$|I - \lambda A| \neq 0$$

The values of λ for which

$$|I - \lambda A| = 0$$

are called the eigenvalues of the integral equation. If $\lambda_1, \lambda_2, \ldots, \lambda_n$ are such eigenvalues, and if λ is chosen to be equal to one of the eigenvalues of the integral equation, the latter has a solution only if

$$\rho(I - \lambda_i A) = \rho(I - \lambda_i A \;\vdots\; b), \quad i = 1, \ldots, n$$

We therefore conclude that the integral equation

$$x(t) = \lambda \int_a^b k(\xi, t)x(\xi)\, d\xi$$

always has a solution, provided that λ is chosen to be equal to one of the eigenvalues $\lambda_1, \lambda_2, \ldots, \lambda_n$. In this case $x_1, \ldots, x_n$ will have more than one solution depending on $\rho(I - \lambda_i A)$.

Example. Solve the integral equation

$$x(t) = e^t + \lambda \int_0^{10} t\xi x(\xi)\, d\xi$$

Here the kernel is degenerate and we have

$$f(t) = e^t, \quad k(t, \xi) = t\xi, \quad q(t) = t, \quad r(\xi) = \xi$$

Therefore we solve the equation

$$(1 - \lambda a)x = b$$

where

$$a = \int_0^{10} \xi^2\, d\xi = \frac{1000}{3}, \quad b = \int_0^{10} \xi e^{\xi}\, d\xi = 10e^{10} - e^{10} + 1$$

and obtain

$$x = \frac{9e^{10} + 1}{1 - \dfrac{1000\lambda}{3}}$$

Hence the integral equation has a solution except for the case when $\lambda = \dfrac{3}{1000}$. The solution is therefore given by:

$$x(t) = e^t + \lambda t \frac{9e^{10} + 1}{1 - \dfrac{1000\lambda}{3}}$$

Example. Solve the integral equation

$$x(t) = t^2 + \int_0^1 \sin(t\xi)x(\xi)\, d\xi$$

Although the kernel is non-degenerate, it can be expanded in a Taylor's series as follows:

$$\sin(t\xi) = t\xi - \frac{t^3\xi^3}{3!} + \cdots$$

Therefore we conclude that there exists an infinite number of eigenvalues, but taking the first approximation we obtain

$$a = \int_0^1 \xi^2 \, d\xi = \tfrac{1}{3}$$

$$b = \int_0^1 \xi\xi^2 \, d\xi = \tfrac{1}{4}$$

Hence

$$(1 - \tfrac{1}{3})x = \tfrac{1}{4}$$

giving

$$x = \frac{3}{8}$$

and

$$x(t) = t^2 + \frac{3}{8} t$$

Taking the second approximation we obtain

$$A = \begin{bmatrix} \int_0^1 \xi \, d\xi^2 & -\int_0^1 \xi\xi^3 \, d\xi \\ \int_0^1 \frac{\xi^3\xi}{3!} \, d\xi & -\int_0^1 \frac{\xi^3\xi^3}{3!} \, d\xi \end{bmatrix} = \begin{bmatrix} \frac{1}{3} & -\frac{1}{5} \\ \frac{1}{30} & \frac{-1}{42} \end{bmatrix}$$

$$b = \begin{bmatrix} \int_0^1 \xi\xi^2 \, d\xi \\ \int_0^1 \frac{\xi^2\xi^3}{3!} \, d\xi \end{bmatrix} = \begin{bmatrix} \frac{1}{4} \\ \frac{1}{36} \end{bmatrix}$$

and solving

$$\begin{bmatrix} \frac{2}{3} & \frac{1}{5} \\ -\frac{1}{30} & \frac{43}{42} \end{bmatrix} \begin{bmatrix} x_1 \\ x_2 \end{bmatrix} = \begin{bmatrix} \frac{1}{4} \\ \frac{1}{36} \end{bmatrix}$$

we obtain approximately

$$x_1 = \frac{3}{8}, \quad x_2 = \frac{3}{80}$$

Hence

$$x(t) = t^2 + \frac{3}{8}t - \frac{3}{80}\frac{t^3}{6}$$

which is an approximate solution. As we have seen, if the kernel is non-degenerate, we have an infinite number of eigenvalues; this is why we approximate to two or three terms only. Other methods for treating integral equations with non-degenerate kernels will be given later.

Now we turn our attention to integral equations with symmetric kernels, since they deserve a treatment of their own. Let the integral equation under investigation be

$$x(t) = \lambda \int_a^b k(\xi, t)x(\xi)\,\mathrm{d}\xi$$

where

$$k(\xi, t) = k(t, \xi)$$

The above integral equation has always a solution corresponding to each eigenvalue of the equation. Let λ_i and λ_j be two such eigenvalues of the equation and correspondingly

$$x_i(t) = \lambda_i \int_a^b k(\xi, t)x_i(\xi)\,\mathrm{d}\xi$$

$$x_j(t) = \lambda_j \int_a^b k(\xi, t)x_j(\xi)\,\mathrm{d}\xi$$

Multiplying by $x_j(t)$ the first equation and integrating, we get:

$$\int_a^b x_i(t)x_j(t)\,\mathrm{d}t = \lambda_i \int_a^b x_j(t) \int_a^b k(\xi, t)x_i(\xi)\,\mathrm{d}\xi\,\mathrm{d}t$$

$$= \lambda_i \int_a^b x_i(\xi) \int_a^b k(\xi, t)x_j(t)\,\mathrm{d}t\,\mathrm{d}\xi,$$

and as the kernel is symmetric we obtain

$$\int_a^b x_i(t)x_j(t)\,dt = \lambda_i \int_a^b x_i(\xi) \int_a^b k(t,\xi)x_j(t)\,dt\,d\xi$$

$$= \frac{\lambda_i}{\lambda_j} \int_a^b x_i(\xi)x_j(\xi)\,d\xi$$

since $\lambda \neq 0$; otherwise we obtain a trivial solution for $x(t)$:

$$(\lambda_j - \lambda_i) \int_a^b x_i(\xi)x_j(\xi)\,d\xi = 0$$

Therefore we have established the following theorem:

THEOREM. *Eigenfunctions* of integral equations with symmetric kernels corresponding to distinct eigenvalues are orthogonal in the interval $[a, b]$. A second theorem is:

THEOREM. The eigenvalues of integral equations with symmetric kernels are all real.

Proof. If λ_i is an eigenvalue, so is $\bar{\lambda}_i$; hence

$$(\lambda_i - \bar{\lambda}_i) \int_a^b x_i(\xi)\bar{x}_i(\xi)\,d\xi = (\lambda_i - \bar{\lambda}_i) \int_a^b |x_i(\xi)|^2\,d\xi = 0$$

Therefore

$$\lambda_i = \bar{\lambda}_i$$

Next we consider a method to obtain the eigenvalues and eigenfunctions of integral equations with symmetric kernel. The method is very similar to the power method of Von Mises, which we applied before to Hermitian and real symmetric matrices. Let

$$x(t) = \lambda \int_a^b k(\xi, t)x(\xi)\,d\xi$$

where $k(\xi, t)$ is symmetric; choose a function $\theta(t)$ and write

$$\theta(t) = \sum_{n=1}^{\infty} a_n \Phi_n(t)$$

where $\Phi_1(t), \Phi_2(t), \ldots$ are the eigenfunctions of the integral equation. Multiplying by $k(\xi, t)$ and integrating several times, we obtain:

$$\theta_1(\xi) = \int_a^b k(\xi, t)\theta(t)\,\mathrm{d}t = \sum_{n=1}^{\infty} a_n \int_a^b k(\xi, t)\Phi_n(t)\,\mathrm{d}t = \sum_{n=1}^{\infty} \frac{a_n \Phi_n(\xi)}{\lambda_n}$$

$$\theta_2(\xi) = \int_a^b k(\xi, t)\theta_1(t)\,\mathrm{d}t = \sum_{n=1}^{\infty} a_n \int_a^b \frac{k(\xi, t)}{\lambda_n}\Phi_n(t)\,\mathrm{d}t = \sum_{n=1}^{\infty} \frac{a_n \Phi_n(\xi)}{\lambda_n^2}$$

$$\vdots$$

$$\theta_m(\xi) = \int_a^b k(\xi, t)\theta_{m-1}(t)\,\mathrm{d}t = \sum_{n=1}^{\infty} a_n \int_a^b \frac{k(\xi, t)}{\lambda_n^{m-1}}\Phi_n(t)\,\mathrm{d}t = \sum_{n=1}^{\infty} \frac{a_n \Phi_n(\xi)}{\lambda_n^m}$$

$$\theta_{m+1}(\xi) = \int_a^b k(\xi, t)\theta_m(t)\,\mathrm{d}t = \sum_{n=1}^{\infty} a_n \int_a^b \frac{k(\xi, t)}{\lambda_n^{m}}\Phi_n(t)\,\mathrm{d}t = \sum_{n=1}^{\infty} \frac{a_n \Phi_n(\xi)}{\lambda_n^{m+1}}$$

The reader will realize that as m becomes very large

$$\frac{\theta_m(\xi)}{\theta_{m+1}(\xi)} \approx \check{\lambda}_j$$

and to obtain $\Phi_j(t)$ corresponding to the minimum eigenvalues $\check{\lambda}_n$ we use

$$\theta_m(t) \approx \frac{a_j \Phi_j(t)}{\check{\lambda}_j^m}$$

i.e.

$$\Phi_j(t) = \frac{\check{\lambda}_j^m}{a_j}\theta_m(t) = \alpha\theta_m(t)$$

where α is chosen such that $\| \Phi_j(t) \| = 1$, i.e. such that

$$\int_a^b \Phi_j(t)\Phi_j(t)\,\mathrm{d}t = 1 = \alpha^2 \int_a^b \theta_m(t)\theta_m(t)\,\mathrm{d}t$$

giving

$$\Phi_j(t) = \frac{\theta_m(t)}{\sqrt{\int_a^b \theta_m(t)\theta_m(t)\,\mathrm{d}t}}$$

To carry the result a bit further, we use a new kernel

$$\tilde{k}(\xi, t) = k(\xi, t) - \frac{\Phi_j(t)}{\check{\lambda}}\Phi_j(\xi)$$

With this kernel the integral equation will have the same eigenvalues except that the smallest $\check{\lambda}$ is transferred to infinity, and it will also have the same eigenfunctions, as the reader can show for himself. So the power method can be reapplied to the new equation of kernel $\tilde{k}(\xi, t)$, giving us a set of eigenvalues $\lambda_1, \lambda_2, \ldots$ and eigenfunctions $\Phi_1(t), \Phi_2(t) \ldots$.

Now we look at the non-homogeneous equation

$$x(t) = f(t) + \lambda \int_a^b k(\xi, t) x(\xi) \, d\xi$$

and try to solve it for $x(t)$. It can be written in the form

$$x(t) = f(t) + \sum_{n=1}^{\infty} a_n \Phi_n(t)$$

where the set of eigenfunctions $\Phi_1(t), \Phi_2(t), \ldots$ are orthonormalized, and satisfy

$$\Phi_n(t) = \lambda_n \int_a^b k(\xi, t) \Phi_n(\xi) \, d\xi$$

Substituting in the integral equation, we obtain

$$f(t) + \sum_{n=1}^{\infty} a_n \Phi_n(t) = f(t) + \lambda \int_a^b k(\xi, t) \left(f(\xi) + \sum_{n=1}^{\infty} a_n \Phi_n(\xi) \right) d\xi$$

Multiplying by $\Phi_j(t)$ and integrating over the interval $[a, b]$ we get

$$a_j = \lambda \int_a^b \Phi_j(t) \int_a^b k(\xi, t) \left(f(\xi) + \sum_{n=1}^{\infty} a_n \Phi_n(\xi) \right) d\xi \, dt$$

$$= \lambda \int_a^b f(\xi) \int_a^b k(\xi, t) \Phi_j(t) \, dt \, d\xi + \lambda \sum_{n=1}^{\infty} a_n \int_a^b \Phi_n(\xi) \int_a^b k(\xi, t) \Phi_j(t) \, dt \, d\xi$$

$$= \lambda \int_a^b \frac{f(\xi) \Phi_j(\xi) \, d\xi}{\lambda_j} + \lambda \sum_{n=1}^{\infty} a_n \int_a^b \frac{\Phi_n(\xi) \Phi_j(\xi) \, d\xi}{\lambda_j}$$

$$= \frac{\lambda}{\lambda_j} f_j + \frac{\lambda}{\lambda_j} a_j$$

Thus

$$a_j = \frac{\lambda f_j}{\lambda_j - \lambda}$$

and the solution $x(t)$ can be obtained from the following relation

$$x(t) = f(t) + \sum_{n=1}^{\infty} \frac{\lambda f_n}{\lambda_n - \lambda} \Phi_n(t)$$

where

$$f_n = \int_a^b f(\xi)\Phi_n(\xi)\, d\xi$$

Example. Solve the integral equation

$$x(t) = 1 + \lambda \int_0^{\pi} \sin(\xi + t)x(\xi)\, d\xi$$

The eigenvalues and orthonormalized eigenfunctions of the homogeneous equation are obtained by iteration to be

$$\lambda_1 = \frac{2}{\pi}, \quad \lambda_2 = -\frac{2}{\pi}.$$

$$\Phi_1 = \frac{1}{\sqrt{\pi}}(\cos t + \sin t), \; \Phi_2 = \frac{-1}{\sqrt{\pi}}(\cos t - \sin t)$$

Therefore

$$f_1 = \int_0^{\pi} \Phi_1\, dt = \frac{2}{\sqrt{\pi}}$$

$$f_2 = \int_0^{\pi} \Phi_2\, dt = \frac{2}{\sqrt{\pi}}$$

Hence the solution is

$$x(t) = 1 + \frac{\lambda}{\frac{2}{\pi} - \lambda} \cdot \frac{2}{\sqrt{\pi}} \cdot \frac{1}{\sqrt{\pi}}(\cos t + \sin t) + \frac{\lambda}{-\frac{2}{\pi} - \lambda} \cdot \frac{2}{\sqrt{\pi}} \cdot \frac{-1}{\sqrt{\pi}}(\cos t - \sin t)$$

$$= 1 + \frac{8\lambda \cos t + 4\pi\lambda^2 \sin t}{4 - \pi^2\lambda^2}$$

Exercises 4.11

1. Solve the integral equation $x(t) = t + \lambda \int_0^1 \cos(\xi t)x(\xi)\, d\xi$.
2. Obtain the eigenvalues and eigenfunctions of the following equations:

(i) $x(t) = \lambda \int_0^1 (3\xi t + \sin t \sin \xi) x(\xi) \, d\xi$

(ii) $x(t) = \lambda \int_0^\pi \cos(t + \xi) x(\xi) \, d\xi$

3. Show that if the kernel of the integral equation $x(t) = \lambda \int_a^b k(\xi, t) x(\xi) \, d\xi$ is such that $k(\xi, t) = -k(t, \xi)$, [$k(\xi, t)$ is skew-symmetric], the eigenvalues of the integral equation are pure imaginary and two eigenfunctions corresponding to two distinct eigenvalues are orthogonal in the interval $[a, b]$. Hence or otherwise solve the integral equation

$$x(t) = \lambda \int_0^\pi \sin(t - \xi) x(\xi) \, d\xi$$

4. The integral equation $x(t) = f(t) + \lambda \int_a^b k(\xi, t) x(\xi) \, d\xi$ can be solved by iteration, i.e. by choosing a guess solution $x^0(t)$ and substituting it in the right-hand side to calculate $x^1(t)$, then substituting similarly $x^1(t)$ to obtain $x^2(t)$, and so forth. Show that the solution converges if $|\lambda| m |b - a| < 1$, where $m = \max_{a \leqslant t, \xi \leqslant b} k(\xi, t)$. Hence solve by iteration the integral equation $x(t) = 1 + \lambda \int_0^1 (1 - 3t\xi) x(\xi) \, d\xi$ to obtain

$$x(t) = (1 + \lambda(1 - (3/2)t)) \left(1 + \frac{\lambda^2}{4} + \frac{\lambda^4}{16} + \cdots \right)$$

5. The numerical solution of the integral equation $x(t) = f(t) + \lambda \int_a^b k(\xi, t) x(\xi) \, d\xi$ is obtained by using one of the numerical methods of integration. For example the integral equation can be approximated as follows:

$$x(t) = f(t) + \lambda \sum_{j=1}^{n} \theta_j k(t, \xi_j) x(\xi_j)$$

where

$$\left.\begin{aligned} \theta_j &= \frac{b - a}{2n}, \quad j = 1, n \\ \theta_j &= \frac{b - a}{n}, \quad j = 2, \ldots, n - 1 \end{aligned}\right\} \text{Trapezoidal rule}$$

or for three points

$$\theta_j = \frac{b - a}{n} [1, 4, 1] \quad \text{Simpson's rule}$$

At point i of the t axis we obtain:

$$x(t_i) = f(t_i) + \lambda \sum_{j=1}^{n} \theta_j k(t_i, \xi_j) x(\xi_j)$$

i.e.

$$x_i = f_i + \lambda \sum_{j=1}^{n} \theta_j k_{ij} x_j, \quad i = 1, 2, \ldots, n$$

in other words

$$(I - \lambda K\theta)x = f$$

where

$$x^T = [x_1, x_2, \ldots, x_n], \quad f^T = [f_1, f_2, \ldots, f_n], \quad \theta = \mathrm{diag}(\theta_1, \theta_2, \ldots, \theta_n)$$

Hence solve numerically the integral equation

$$x(t) = 1 + \int_0^{\frac{\pi}{2}} \sin(\xi t) x(\xi)\, \mathrm{d}\xi$$

For a survey of the relevant numerical methods, the reader is referred to Delves and Walsh (1974).

4.12. Application to function minimization

To minimize a continuously differentiable function of many variables $f(x_1, x_2, \ldots, x_n)$, we require the calculation of a vector $\tilde{x}_1, \tilde{x}_2, \ldots, \tilde{x}_n$ at which f is minimum. If f is sought to be maximum, then one can minimize $-f$.

The point at which f attains a maximum or a minimum is called a local maximum or a local minimum of the function f. If the function is unconstrained, in other words if it is studied in the open interval $(-\infty, \infty)$, the global maximum or minimum of the function will be identical to one of the local maxima or minima. Therefore we direct ourselves to determine the local minima or maxima of f.

Assume $\tilde{x}$ be a point at which f attains a local minimum or maximum. Then, as f is a continuously differentiable function, it can be expanded by Taylor's series around $\tilde{x}$ as follows:

$$f(x_1, \ldots, x_n) = f(\tilde{x}_1, \tilde{x}_2, \ldots, \tilde{x}_n) + (x_1 - \tilde{x}_1) \left.\frac{\partial f}{\partial x_1}\right|_{\tilde{x}} + (x_2 - \tilde{x}_2) \left.\frac{\partial f}{\partial x_2}\right|_{\tilde{x}} + \cdots +$$

$$+ (x_n - \tilde{x}_n) \left.\frac{\partial f}{\partial x_n}\right|_{\tilde{x}} + \frac{1}{2!}(x_1 - \tilde{x}_1)^2 \left.\frac{\partial^2 f}{\partial x_1^2}\right|_{\tilde{x}} +$$

$$+ \frac{1}{2!}(x_2 - \tilde{x}_2)^2 \left.\frac{\partial^2 f}{\partial x_2^2}\right|_{\tilde{x}} + (x_1 - \tilde{x}_1)(x_2 - \tilde{x}_2) \left.\frac{\partial^2 f}{\partial x_1 \partial x_2}\right|_{\tilde{x}} + \cdots$$

The above expansion can be found in Sokolnikoff (1939), p. 317. It can also be put in a more convenient form as follows:

$$f(x) = f(\tilde{x}) + g^T|_{\tilde{x}}\,(x - \tilde{x}) + \tfrac{1}{2}(x - \tilde{x})^T H|_{\tilde{x}}\,(x - \tilde{x}) + \cdots$$

where g is a vector containing the first partial derivatives and is given by

$$g^T = \left[\frac{\partial f}{\partial x_1}, \frac{\partial f}{\partial x_2}, \ldots, \frac{\partial f}{\partial x_n}\right]$$

H is called the *Hessian* matrix containing the second partial derivatives and is given by

$$H = \begin{bmatrix} \frac{\partial^2 f}{\partial x_1^2} & \frac{\partial^2 f}{\partial x_1 \partial x_2} & \cdots & \frac{\partial^2 f}{\partial x_1 \partial x_n} \\ \frac{\partial^2 f}{\partial x_2 \partial x_1} & \frac{\partial^2 f}{\partial x_2^2} & \cdots & \\ \vdots & & & \\ \frac{\partial^2 f}{\partial x_n \partial x_1} & & \cdots & \frac{\partial^2 f}{\partial x_n^2} \end{bmatrix}$$

The point $\tilde{x}$ at which f has a local minimum or maximum is given by

$$g|_{\tilde{x}} = 0$$

which amounts to n equations to solve for $\tilde{x}_1, \tilde{x}_2, \ldots, \tilde{x}_n$. The reason why $g = 0$ at a local maximum or minimum is that for a virtual displacement of the function f around $\tilde{x}$ we have

$$\delta f(x) = g^T|_{\tilde{x}}\,\delta(x)$$

$\tilde{x}$ is defined as the point at which

$$\delta f(x) = 0$$

while

$$\delta(x) \neq 0$$

Hence we obtain

$$g^T|_{\tilde{x}}\,\delta x = 0$$

But as the variables $\delta x_1, \delta x_2, \ldots, \delta x_n$ can be chosen arbitrarily, we must have

$$\left.\frac{\partial f}{\partial x_1}\right|_{\tilde{x}} = \left.\frac{\partial f}{\partial x_2}\right|_{\tilde{x}} = \cdots = \left.\frac{\partial f}{\partial x_n}\right|_{\tilde{x}} = 0$$

Once $\tilde{x}$ is determined, it remains to determine whether $\tilde{x}$ is a point of local maximum or local minimum for f. Writing Taylor's expansion again with $g|_{\tilde{x}} = 0$ gives

$$f(x) = f(\tilde{x}) + \tfrac{1}{2}(x - \tilde{x})^T H|_{\tilde{x}} (x - \tilde{x}) + \cdots$$

Now if $\tilde{x}$ is a point of local maximum, we have

$$f(x) < f(\tilde{x})$$

giving for small enough $(x - \tilde{x})$,

$$(x - \tilde{x})^T H|_{\tilde{x}} (x - \tilde{x}) < 0, \quad \forall (x - \tilde{x}) \neq 0$$

i.e. $H|_{\tilde{x}}$ is a negative definite matrix. Instead if $\tilde{x}$ is a point of local minimum, then

$$f(x) > f(\tilde{x})$$

giving

$$(x - \tilde{x})^T H|_{\tilde{x}} (x - \tilde{x}) > 0, \quad \forall (x - \tilde{x}) \neq 0$$

i.e. $H|_{\tilde{x}}$ is positive definite.

Example. Find the condition for $f(x, y)$ to have a local maximum or minimum in the interval $(-\infty, \infty)$.

The point of local maximum or minimum (x, y) is given by

$$\frac{\partial f}{\partial x} = \frac{\partial f}{\partial y} = 0, \quad \Rightarrow \tilde{x}, \tilde{y}$$

However if $(\tilde{x}, \tilde{y})$ is a point of local maximum the matrix

$$\begin{bmatrix} \left.\dfrac{\partial^2 f}{\partial x^2}\right|_{\tilde{x},\tilde{y}} & \left.\dfrac{\partial^2 f}{\partial x \partial y}\right|_{\tilde{x},\tilde{y}} \\ \left.\dfrac{\partial^2 f}{\partial y \partial x}\right|_{\tilde{x},\tilde{y}} & \left.\dfrac{\partial^2 f}{\partial y^2}\right|_{\tilde{x},\tilde{y}} \end{bmatrix}$$

is negative definite, i.e. the condition for $(\tilde{x}, \tilde{y})$ to be a point of local maximum is

$$\left.\frac{\partial^2 f}{\partial x^2}\right|_{\tilde{x},\tilde{y}} < 0, \qquad \begin{vmatrix} \left.\dfrac{\partial^2 f}{\partial x^2}\right|_{\tilde{x},\tilde{y}} & \left.\dfrac{\partial^2 f}{\partial x \partial y}\right|_{\tilde{x},\tilde{y}} \\ \left.\dfrac{\partial^2 f}{\partial y \partial x}\right|_{\tilde{x},\tilde{y}} & \left.\dfrac{\partial^2 f}{\partial y^2}\right|_{\tilde{x},\tilde{y}} \end{vmatrix} > 0$$

And for $(\tilde{x}, \tilde{y})$ to be a point of local minimum we must have $H|_{\tilde{x},\tilde{y}}$ positive definite, i.e.

$$\left.\frac{\partial^2 f}{\partial x^2}\right|_{\tilde{x},\tilde{y}} > 0 \qquad \begin{vmatrix} \left.\frac{\partial^2 f}{\partial x^2}\right|_{\tilde{x},\tilde{y}} & \left.\frac{\partial^2 f}{\partial x \partial y}\right|_{\tilde{x},\tilde{y}} \\ \left.\frac{\partial^2 f}{\partial y \partial x}\right|_{\tilde{x},\tilde{y}} & \left.\frac{\partial^2 f}{\partial y^2}\right|_{\tilde{x},\tilde{y}} \end{vmatrix} > 0$$

From what has been said, we can obtain a maximum or a minimum point of the function $f(x_1, x_2, \ldots, x_n)$ by setting $g = 0$; but this is not easy after that, for setting $g = 0$ may generate a set of n nonlinear simultaneous equations in $\tilde{x}_1, \tilde{x}_2, \ldots, \tilde{x}_n$, whose solution is by no means trivial. So this method of obtaining the maximum or minimum of a function can be applied only when $f(x_1, x_2, \ldots, x_n)$ is a quadratic function, so that $g = 0$ generates a set of n linear simultaneous equations easy to solve. An example of that has been seen when we required to solve

$$Ax = b$$

with A of order (m, n) and $m > n$ (see Section 3.9c). The problem becomes that of finding x which minimizes $\| Ax - b \|_2^2$; this is a quadratic function in x, and hence the vector g containing the first-order partial derivatives is linear in x. The solution was found to be that of

$$A^T A x = A^T b$$

Note that the Hessian of f is $A^T A$, which is seen to be a constant positive definite matrix when $\rho(A) = n$, meaning that the problem is posed as a minimization problem. Another example is the least-squares method for fitting the straight line $y = ax + b$ to the set of points $(x_1, y_1), (x_2, y_2), \ldots, (x_n, y_n)$. The problem becomes that of finding the constants a and b which minimize the function

$$f(a, b) = \sum_{i=1}^{n} (y_i - ax_i - b)^2$$

They are given by

$$\frac{\partial f}{\partial a} = 0 = 2 \sum_{i=1}^{n} (y_i - ax_i - b)(-x_i)$$

and

$$\frac{\partial f}{\partial b} = 0 = 2 \sum_{i=1}^{n} (y_i - ax_i - b)(-1)$$

The above equations can be organized in the convenient form

$$\begin{bmatrix} n & \sum_{i=1}^{n} x_i \\ \sum_{i=1}^{n} x_i & \sum_{i=1}^{n} x_i^2 \end{bmatrix} \begin{bmatrix} b \\ a \end{bmatrix} = \begin{bmatrix} \sum_{i=1}^{n} y_i \\ \sum_{i=1}^{n} y_i x_i \end{bmatrix}$$

which is the same solution obtained before in Exercise 3.10.14 by substituting the points $(x_1, y_1), (x_2, y_2), \ldots, (x_n y_n)$ in the equation of the straight line, giving n equations in the two unknowns a and b. The reader will notice that the matrix of the right-hand side is nonsingular because

$$n \sum_{i=1}^{n} x_i^2 > \left(\sum_{i=1}^{n} x_i \right)^2$$

a result which can be obtained using the Holder inequality explained in Appendix 1. Also we can easily show that this same matrix is the Hessian of the function $f(a, b)$; this being positive definite means that we are again dealing with a minimization problem.

Many other nonlinear functions which seem difficult to deal with, except by nonlinear techniques, can be dealt with using the same procedure explained above if they can be formulated as a quadratic objective function. For example one can fit a set of points to the curve $y = ae^{bx}$. The procedure is to take the logarithm of both sides to formulate the problem as follows:

$$\min_{a,b} \sum_{i=1}^{n} (\log y_i - \log a - bx_i)^2$$

If $f(x_1, x_2, \ldots, x_n)$ is not quadratic, the above methods fail, for $g = 0$ generates a set of n nonlinear simultaneous equations. The procedure will then be to choose a guess point x^0 and improve on it until the solution is reached. The procedure is as follows. Expand $f(x)$ around x^0 by Taylor's series:

$$f(x) = f(x^0) + g^T \big|_{x^0} (x - x^0) + \tfrac{1}{2}(x - x^0)^T H \big|_{x^0} (x - x^0) + 0((x - x^0)^3)$$

Now $\bar{x}$ is defined as the point at which

$$\left. \frac{\partial f}{\partial x} \right|_{\bar{x}} = 0$$

Differentiating the Taylor's expansion w.r.t. x or $(x - x^0)$ gives

$$0 \approx g \big|_{x^0} + H \big|_{x^0} (\bar{x} - x^0)$$

where the third term is neglected if x^0 is rightly chosen near $\bar{x}$. Hence

$$\bar{x} \approx x^0 - H^{-1} \big|_{x^0} g \big|_{x^0}$$

where the vector $y = H^{-1} g$ is obtained by solving the linear equations

$$Hy = g$$

Because the above correction for $\bar{x}$ is only approximate, the solution $\bar{x}$ can be improved by iteration.

The convergence of the above method is guaranteed if H after every iteration is found positive definite for a minimization problem or negative definite for a maximization problem. For example, for a minimization problem,

$$f(\tilde{x}) = f(x^0) + g^T|_{x^0}(\tilde{x} - x^0) + \tfrac{1}{2}(\tilde{x} - x^0)^T H|_{x^0}(x - x^0)$$

$$= f(x^0) - g^T|_{x^0} H^{-1}|_{x^0} g|_{x^0} + \tfrac{1}{2} g^T|_{x^0} H^{-1}|_{x^0} g|_{x^0}$$

$$= f(x^0) - \tfrac{1}{2} g^T|_{x^0} H^{-1}|_{x^0} g|_{x^0}$$

and since $H|_{x^0}$ is positive definite, $H^{-1}|_{x^0}$ is also positive definite, as the reader can show. Hence

$$g^T|_{x^0} H^{-1}|_{x^0} g|_{x^0} > 0$$

From this we obtain

$$f(\tilde{x}) < f(x^0)$$

which proves the convergence of the method.

One main disadvantage of using the above method for calculating $\tilde{x}$ is that H should be calculated at each iteration and found also positive definite. It requires also the solution of a set of linear equations.

Among older methods which require the evaluation of H but which avoid the solution of linear equations are the *descent methods*. An illustration of these follows. If at the ith iteration, a correction x^{i+1} is required for x^i, such that

$$x^{i+1} = x^i + \alpha_i d^i,$$

the direction of search d^i is chosen to make

$$f(x^{i+1}) < f(x^i)$$

As for the magnitude of search α_i, it is obtained from minimizing $f(x^i + \alpha_i d^i)$.

Expanding $f(x^{i+1})$ around x^i, we obtain

$$f(x^{i+1}) \approx f(x^i) + \alpha_i d^{iT} g|_{x^i} < f(x^i)$$

i.e. d^i should be chosen such that

$$d^{iT} g|_{x^i} < 0$$

As for α_i, it is obtained by minimizing the function

$$f(x^i + \alpha_i d^i) = f(x^i) + \alpha_i d^{iT} g|_{x^i} + \tfrac{1}{2}\alpha_i^2 d^i H|_{x^i} d^i + 0(\alpha_i^3)$$

which gives approximately

$$\alpha_i = -\frac{\langle d^i, g|_{x^i}\rangle}{\langle d^i, H|_{x^i} d^i\rangle}$$

The function $f(x^{i+1})$ will then be less than $f(x^i)$ if $H|_{x^i}$ is positive definite. Note that α_i could be well obtained from the geometric relation

$$0 = \langle d^i, g|_{x^{i+1}}\rangle$$

$$= \langle d^i, g|_{x^i} + \alpha_i H|_{x^i} d^i\rangle$$

as demonstrated in the figure below

As the choice of d^i is not unique, since it should only satisfy

$$d^{iT} g|_{x^i} < 0$$

then in order not to leave it to a matter of preference, d^i is chosen simply as the downhill gradient vector:

$$d^i = -g|_{x^i}$$

i.e. the direction is chosen in the direction of steepest descent, and the method is then called the *steepest descent method.*

In practice, the steepest descent method usually improves $f(x)$ rapidly on the first few iterations and then gives rise to oscillatory progress and becomes unsatisfactory. The reason, as Fletcher (1969), p. 2, points out, lies in the failure of the theory to represent adequately functions with minima. The only functions for which the steepest descent method properly works are those with spherical contours.

Another popular method which does not rely on solving a set of linear equations is the method of conjugate directions. A set of vectors $d^1, d^2, \ldots, d^n$ is called a set of conjugate directions with respect to a positive definite matrix H, if

$$\langle d^i, H d^j \rangle = 0, \quad i \neq j$$

as was defined in Exercise 4.10.14. From an initial approximation x^1, a direction of search is chosen as

$$d^1 = -g^1$$

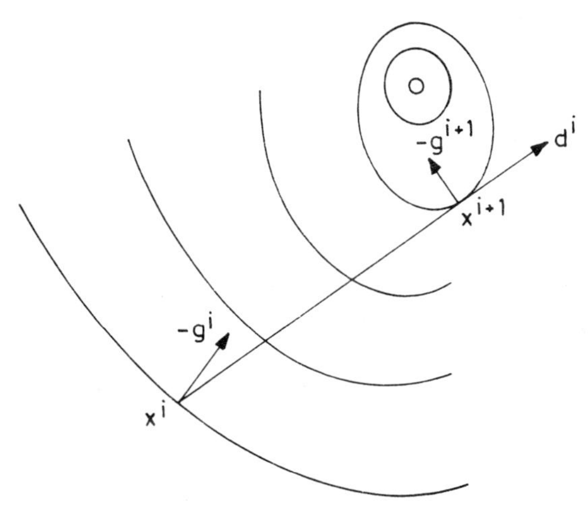

Searching along a line for a minimum on the contour line $f(x)$ = constant and α_1 is obtained as before, to be

$$\alpha_1 = -\frac{\langle d^1, g^1 \rangle}{\langle d^1, H_1 d^1 \rangle} = \frac{\langle g^1, g^1 \rangle}{\langle d^1, H_1 d^1 \rangle}$$

This determines a better approximation x^2 of the solution. The second direction of search is chosen conjugate to d^1 as follows:

$$d^2 = -g^2 + \frac{\langle g^2, g^2 \rangle}{\langle g^1, g^1 \rangle} d^1$$

The reason why d^2 is conjugate to d^1 is that

$$\begin{aligned}\langle d^1, H_1 d^2 \rangle &= -\langle d^1, H_1 g^2 \rangle + \frac{\langle g^2, g^2 \rangle}{\langle g^1, g^1 \rangle} \langle d^1, H_1 d^1 \rangle \\ &= -\left\langle \left(\frac{g^2 - g^1}{\alpha_1} \right), g^2 \right\rangle + \frac{\langle g^2, g^2 \rangle}{\langle g^1, g^1 \rangle} \langle d^1, H_1 d^1 \rangle \\ &= -\frac{\langle g^2, g^2 \rangle}{\langle g^1, g^1 \rangle} \langle d^1, H_1 d^1 \rangle + \frac{\langle g^1, g^2 \rangle}{\alpha_1} + \frac{\langle g^2, g^2 \rangle}{\langle g^1, g^1 \rangle} \langle d^1, H_1 d^1 \rangle \\ &= 0\end{aligned}$$

since

$$0 = \langle d^1, g^2 \rangle = \langle g^1, g^2 \rangle.$$

And minimizing in the direction of x^3, we obtain

$$\alpha_2 = -\frac{\langle d^2, g^2 \rangle}{\langle d^2, H_2 d^2 \rangle} = \frac{\langle g^2, g^2 \rangle}{\langle d^2, H_2 d^2 \rangle}$$

In general

$$d^{i+1} = -g^{i+1} + \frac{\langle g^{i+1}, g^{i+1} \rangle}{\langle g^i, g^i \rangle} d^i$$

and

$$\alpha_i = \frac{\langle g^i, g^i \rangle}{\langle d^i, H_i d^i \rangle}$$

To show that d^3 is conjugate to d^2 and d^1, the reader will rely mainly on the property

$$\langle g^2, g^3 \rangle = 0$$

The above property holds because g^2 is a linear combination of d^2 and d^1. But d^2 is orthogonal on g^3, and so is d^1, for

$$\begin{aligned}\langle g^3, d^1 \rangle &= \langle g^2 + \alpha_2 H_2 d^2, d^1 \rangle \\ &= \langle g^2, d^1 \rangle + \alpha_2 \langle d^2, H_2 d^1 \rangle \\ &= 0\end{aligned}$$

In working out the above proof, it was assumed that the directions $d^1, d^2, \ldots, d^n$ are a set of conjugate directions with respect to a varying positive definite Hessian H. This is why the conjugate directions method works ideally for the case when $f(x)$ is quadratic, i.e. in the form

$$f(x) = a + b^T x + \tfrac{1}{2}\langle x, Ax \rangle$$

where A is positive definite. In this case we reach the solution in exactly n iterative steps. The reason is that g^{n+1} is orthogonal on $d^1, \ldots, d^n$; and the latter being linearly independent, g^{n+1} must be equal to zero (see Exercise 3.5.18), which is exactly the condition of a minimum.

One interesting observation about the conjugate directions method reveals that it can be applied to the solution of linear equations when the matrix of coefficients is positive definite, since the minimum of the above quadratic function, given by

$$\frac{\partial}{\partial x} f(x) = b + Ax = 0$$

is synonymous with the solution of the set of linear simultaneous equations

$$Ax = -b$$

For computational aspects of this problem, the reader is referred to Stoer and Bulirsch (1980), p. 572, who describe the method of Hestenes and Stiefel (1952), for treating the problem. As for the application of the method of conjugate directions to general nonlinear unconstrained functions, the reader is referred to Fletcher and Reeves (1964).

The descent methods and conjugate directions method have enabled one to avoid the solution of a set of linear simultaneous equations which is necessary if one uses the formula

$$x^{i+1} = x^i - H_i^{-1} g^i$$

So if a representation of H^{-1} can be achieved, it is sure enough that such a method will supersede the previous methods; since for example for a quadratic function the new method will reach the solution in one step against n steps for the method of conjugate directions. The reader can check that H^{-1} can be represented in the form

$$H^{-1} = \sum \frac{d^i d^{iT}}{\langle d^i, H_i d^i \rangle}$$

The best known implementation of this formula has been the DFP method devised by Davidon (1959) and by Fletcher and Powell (1963). Not only has this method the advantage of rapid convergence, but it can also be used when only the gradient vector g^i is known. A situation like this is frequently encountered in physical problems, where sometimes the Hessian matrix H cannot be easily calculated. Fletcher and Powell suggested a correction for the inverse Z of the Hessian in the form

$$Z_{i+1} = Z_i + \frac{\partial^i \partial^{iT}}{\langle \partial^i, Z_i\, \partial^i \rangle} - \frac{Z_i p^i p^{iT} Z_i}{\langle p^i, Z_i p^i \rangle}$$

where

$$\partial^i = x^{i+1} - x^i$$

$$p^i = g^{i+1} - g^i$$

and where Z_1 can be chosen to be a unit matrix unless an approximate value for the inverse of H is known. The reader can check that if Z_1 is positive definite, then all subsequent Z_i are also positive definite.

However in some problems, one finds that H is always positive definite and need not be calculated at all, as in problems treated by the generalized least squares method. If f is a summation of squares of nonlinear functions like the following:

$$f = \sum_{j=1}^{m} \Phi_j^2\,(x_1, x_2, \ldots, x_n)$$

then

$$g_k = \frac{\partial f}{\partial x_k} = 2\sum_{j=1}^{m} \Phi_j\,(x_1, x_2, \ldots, x_n)\,\frac{\partial \Phi_j}{\partial x_k}$$

and

$$H_{ik} = \frac{\partial^2 f}{\partial x_i \partial x_k} = 2\sum_{j=1}^{m} \Phi_j\,(x_1, x_2, \ldots, x_n)\,\frac{\partial^2 \Phi_j}{\partial x_i \partial x_k} + 2\sum_{j=1}^{m} \frac{\partial \Phi_j}{\partial x_i} \cdot \frac{\partial \Phi_j}{\partial x_k}$$

Neglecting the first term in H_{ik} if x^0 is rightly chosen near $\tilde{x}$ such that $\Phi_j(x^0)$ is small, we obtain

$$g\,|_{x^0} = 2\,J^T\,|_{x^0}\,\Phi\,|_{x^0}$$

and

$$H\,|_{x^0} \approx 2\,J^T\,|_{x^0} J\,|_{x^0}$$

where J is called the *Jacobian* matrix of transformation from Φ into x and is given by

$$J_{m,n} = \begin{bmatrix} \frac{\partial \Phi_1}{\partial x_1} & \frac{\partial \Phi_1}{\partial x_2} & \cdots & \frac{\partial \Phi_1}{\partial x_n} \\ \frac{\partial \Phi_2}{\partial x_1} & \frac{\partial \Phi_2}{\partial x_2} & \cdots & \\ \vdots & & & \\ \frac{\partial \Phi_m}{\partial x_1} & & \cdots & \frac{\partial \Phi_m}{\partial x_n} \end{bmatrix}$$

Φ is the error vector and is given by

$$\Phi^T = [\Phi_1, \Phi_2, \ldots, \Phi_m]$$

Therefore the approximate solution x is obtained from the correction formula

$$\begin{aligned} \tilde{x} &\approx x^0 - H^{-1}|_{x^0} g|_{x^0} \\ &\approx x^0 - (J^T|_{x^0} J|_{x^0})^{-1} J^T|_{x^0} \Phi|_{x^0} \end{aligned}$$

If $n > m$, then $J^T J$ is singular, as the reader can verify. This is why the above equation is altered, see Levenberg (1944) and Marquardt (1963), to be

$$\tilde{x} \approx x^0 - (J^T|_{x^0} J|_{x^0} + \lambda I)^{-1} J^T|_{x^0} \Phi|_{x^0}$$

where λ is a scalar ranging from zero to a very high value. In this case why is the first matrix on the right-hand side positive definite?

But if $m > n$, then the reader can show that $J^T J$ is nonsingular when $\rho(J) = n$. Moreover $J^T J$ is always positive definite; hence the problem is a minimization one. However as the iteration proceeds, the rank of J decreases near the solution especially if f is not zero at its minimum. This is why we find that J becomes singular, near the solution, even if $m = n$. Thus, in all cases, the λ term is added to guarantee the convergence. Only in some cases, the λ term is not needed when the minimum of f is actually equal to zero. In this case, convergence becomes rapid and J does not have the chance to be singular; for the solution is reached quickly before that phenomenon is allowed to happen. However if f is zero at the minimum, in other words if the minimum of f is zero, then

$$\Phi_j|_{\tilde{x}} = 0, \quad \forall_j = 1, 2, \ldots, m$$

since f is a summation of squares of real functions. An important example is to seek the solution of a set of nonlinear simultaneous equations

$$\Phi_j(x_1, x_2, \ldots, x_n) = 0, \quad j = 1, 2, \ldots, n$$

We construct the function

$$f = \sum_{j=1}^{n} \Phi_j^2(x_1, \ldots, x_n)$$

which we seek to minimize, and whose solution is obtained similarly from the correction formula

$$\tilde{x} \approx x^0 - (J^T|_{x^0} J|_{x^0})^{-1} J^T|_{x^0} \Phi|_{x^0}$$

where J is here a square matrix; moreover if $J|_{x^0}$ is nonsingular, then, using Exercise 5.4.9, we obtain Newton's formula of correction

$$\tilde{x} \approx x^0 - J^{-1}|_{x^0} \Phi|_{x^0}$$

Another alternative for solving the above set of nonlinear simultaneous equations, and which amounts to the above formula, is by expanding the equations $\Phi_j|_{\tilde{x}} = 0$ around x^0 using Taylor's series to give

$$0 \approx \Phi(x^0) + J|_{x^0}(\tilde{x} - x^0)$$

from which we obtain the same result. However, as the solution is not guaranteed, not because of the problem of convergence, but because the equations may not have a solution at all, like for example finding the point of intersection of the two curves in the figure shown below, it may be then wiser always to assume the solution

$$\tilde{x} \approx x^0 - (J^T|_{x^0} J|_{x^0} + \lambda I)^{-1} J^T|_{x^0}|_{x^0}$$

In this case a solution is found which is not of course a point of intersection, but is as near as possible to both curves.

In a situation like this, convergence is never reached but the iterations can be ended if the values of the elements of x do not change much, and of course one cannot make sure of that unless the λ term is chosen very large.

In brief, it is always recommended to start the iterations with a large λ and if it is found that convergence is occurring, λ is decreased to accelerate the convergence, until λ is zero, which is of course the ideal case. In other words, if λ is chosen large, there is more possibility of convergence but it will be slow. If λ is small, rapid change in x occurs but the possibility of convergence is scarce; unless of course the the function f is zero at its minimum.

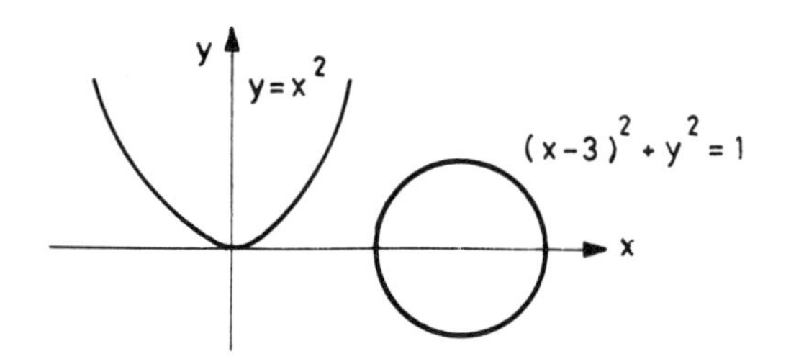

Example. Solve the equations

$$x^2 + y^2 = 5, \quad y = x^3$$

Taking an initial guess

$$x = 1, \quad y = 1$$

we write

$$\Phi^T = [x^2 + y^2 - 5 \quad y - x^3]$$

$$J = \begin{bmatrix} 2x & 2y \\ -3x^2 & 1 \end{bmatrix}$$

First iteration:

$$\begin{bmatrix} x \\ y \end{bmatrix} = \begin{bmatrix} 1 \\ 1 \end{bmatrix} - \begin{bmatrix} 2 & 2 \\ -3 & 1 \end{bmatrix}^{-1} \begin{bmatrix} -3 \\ 0 \end{bmatrix}$$

where the second vector on the right-hand side is obtained by solving the equations

$$\begin{bmatrix} 2 & 2 \\ -3 & 1 \end{bmatrix} \begin{bmatrix} a \\ b \end{bmatrix} = \begin{bmatrix} -3 \\ 0 \end{bmatrix}$$

giving

$$a = -\frac{3}{8}, \quad b = -\frac{9}{8}$$

hence

$$x = 1 + \frac{3}{8} = \frac{11}{8}, \quad y = 1 + \frac{9}{8} = \frac{17}{8}$$

Second iteration:

$$\begin{bmatrix} x \\ y \end{bmatrix} = \begin{bmatrix} \frac{11}{8} \\ \frac{17}{8} \end{bmatrix} - \begin{bmatrix} \frac{11}{4} & \frac{17}{4} \\ -3\left(\frac{11}{8}\right)^2 & 1 \end{bmatrix}^{-1} \begin{bmatrix} \left(\frac{11}{8}\right)^2 + \left(\frac{17}{8}\right)^2 - 5 \\ \frac{17}{8} - \left(\frac{11}{8}\right)^3 \end{bmatrix}$$

Solving for the second vector on the right-hand side, we obtain

$$x = 1.27, \quad y = 1.85$$

The process converges since

$$\| \Phi_0 \| > \| \Phi_1 \| > \| \Phi_2 \|$$

for approximately

$$\Phi_0^T = [-3 \quad 0]$$

$$\Phi_1^T = [1.4 \quad -0.5]$$

$$\Phi_2^T = [0 \quad -0.15]$$

Hence the approximate solution is (1.27, 1.85) and of course the accuracy can be increased by having more iterations.

The generalized least-squares method requires the evaluation of the matrix J. A method due to Barnes (1965) shows how different approximations to J can be set up and updated at each iteration. Both methods still require the solution of a set of linear equations; however this can be avoided by updating an approximation to J^{-1}. This is done by Broyden (1965) and also in the secant method of Wolfe (1959).

The same technique can be applied to fit a nonlinear function of many variables to a set of points. Let there exist a set of points given by $\theta(s_i)$, where s can represent time or frequency, and it is required to fit a nonlinear function $\Phi(x_1, x_2, \ldots, x_n, s)$ to $\theta(s_i)$, i.e. to find $x_1, x_2, \ldots, x_n$ which allow for the best fit. We construct the function

$$f = \sum_{i=1}^{m} (\Phi(x_1, x_2, \ldots, x_n, s_i) - \theta(s_i))^2$$

which we seek to minimize. Applying the same procedure as explained before, we obtain

$$\tilde{x} \approx x^0 - (J^T|_{x^0} J|_{x^0} + \lambda I)^{-1} J^T|_{x^0} (\Phi|_{x^0} - \theta)$$

where

$$\Phi^T|_{x^0} = [\Phi(x^0, s_1), \Phi(x^0, s_2), \ldots, \Phi(x^0, s_m)]$$

$$\theta^T = [\theta(s_1), \theta(s_2), \ldots, \theta(s_m)]$$

$$J_{m,n}|_{x^0} = \begin{bmatrix} \left.\dfrac{\partial\Phi}{\partial x_1}\right|_{(x^0, s_1)} & \left.\dfrac{\partial\Phi}{\partial x_2}\right|_{(x^0, s_1)} & \cdots & \left.\dfrac{\partial\Phi}{\partial x_n}\right|_{(x^0, s_1)} \\ \vdots & & & \\ \left.\dfrac{\partial\Phi}{\partial x_1}\right|_{(x^0, s_m)} & \left.\dfrac{\partial\Phi}{\partial x_2}\right|_{(x^0, s_m)} & \cdots & \left.\dfrac{\partial\Phi}{\partial x_n}\right|_{(x^0, s_m)} \end{bmatrix}$$

The methods discussed above require the explicit calculation of partial derivatives. If the latter are difficult to obtain, we recommend Powell's algorithm (1964). It is based on the observation that if the minimum of a positive definite quadratic form is sought in the direction d from each of two distinct points, then the vector joining the resulting minima is conjugate to d. This is illustrated in the following figure:

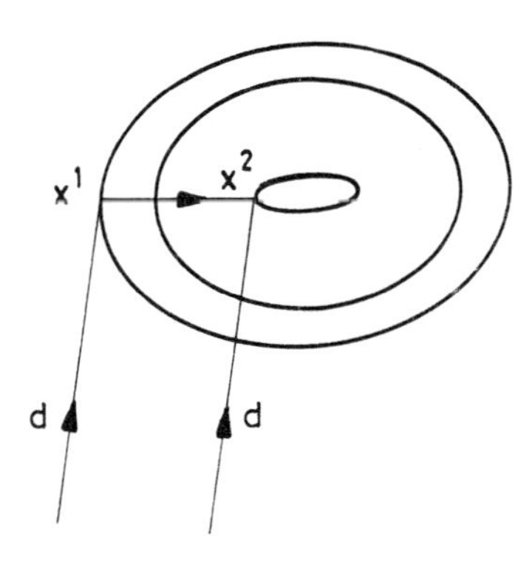

If x^1 and x^2 are two minima, we have:

$$\langle d, Ax^1 + b \rangle = 0$$

and

$$\langle d, Ax^2 + b \rangle = 0$$

Hence, subtracting, we obtain

$$\langle d, A(x^1 - x^2) \rangle = 0$$

Powell's algorithm uses a more elaborate result, namely that if the search for each minimum is made along d conjugate directions, the join of these minima is conjugate to all these directions. For the result of each search depends only on the starting points and not on the order in which the descent steps are carried out. Therefore it can be arranged that any of the d conjugate directions can be used last. The reader interested in function minimization without using derivatives is referred to the review paper by Fletcher (1965), who surveyed available methods. For a general review of the solution of optimization problems, including constrained functions, the reader should see the important paper by Fletcher (1971), in which he also set out a decision tree for the choice of a suitable algorithm.

Exercises 4.12

1. Minimize the function $2x_1^4 + x_2^2 - 4x_1x_2 + 5x_2$.
2. Show that for a general nonlinear function $f(x)$, the method of putting $x = x^0 - H^{-1}|_{x^0} g|_{x^0}$ is itself the same method as that of first putting $g = 0$ and then solving the set of nonlinear equations generated, using the Newton–Raphson method.
3. Solve the equations $x^2 + y^2 + z^2 = 9, x^4 + y^4 = 2, x + y = 1$.
4. Solve the equations $2x^2 + 3y^2 = 7, y = x^4$.
5. Fit the curve $y = x_1^2 + x_2^2 - w^2 x_1$ to the points

w	0	1	2	3	4	5
y	5	4	1	−4	−11	−19

6. Minimize the function $(3x_1^2 - 2x_1x_2 + x_2^2 + x_1)^2 + (2x_1^4 + x_2^2 - 4x_1x_2 + 5x_2 - 10)$
7. For what values of x_1 and x_2 is the quadratic form $(a_{11}x_1 + a_{12}x_2 - b_1)^2 + (a_{21}x_1 + a_{22}x_2 - b)^2$ minimum?
8. Show that the problem of minimizing $\langle x, Bx \rangle + 2\langle x, Ay \rangle + \langle y, By \rangle - 2\langle a, x \rangle - 2\langle b, y \rangle$ is that of solving the equations $Bx + Ay = a$, $A^T x + By = b$.
9. Explain how to deal with the set of linear simultaneous equations $Ax = b$, if A is of order (m, n), $m > n$, and $\rho(A) < n$. *Hint*: consider the positive definite matrix $(A^T A + \lambda I)$; see also Fletcher (1968).
10. Show how to fit the curve $y = a_m x^m + a_{m-1}x^{m-1} + \cdots + a_1 x + a_0$ to a set of points (x_i, y_i), $i = 1, \ldots, n$, $n \geqslant m$.
11. Show how to fit the function $z = ae^{\alpha x + \beta y}$ to a set of points $z_k(x_i, y_j)$, $i = 1, \ldots, m$, $j = 1, \ldots, n$, $k = 1, \ldots, m \times n$.
12. Show for the descent method whose correction formula for $f(x)$ is

$$f(x^{i+1}) = f(x^i) + \alpha_i \langle d^i, g^i \rangle$$

where $\langle d^i, g^i \rangle$ is negative to allow for $f(x^{i+1}) < f(x^i)$, that $\lim_{i \to \infty} g^i = 0$. *Hint*: $f(x^i)$ is monotonic decreasing and bounded below, therefore it is convergent; use the formula

$$f(x^i) = f(x) - \sum_{j=1}^{i} \frac{\langle d^j, g^j \rangle^2}{\langle d^j, H_j d^j \rangle}.$$

13. Show that the rate of convergence of the steepest-descent method $f(x^{i+1})/f(x^i)$ is proportional to $\left(\frac{\hat{\lambda} - \check{\lambda}}{\hat{\lambda} + \check{\lambda}}\right)^2$ with $\hat{\lambda}$ and $\check{\lambda}$ being the largest and smallest eigenvalues of the positive definite Hessian matrix H.
14. Show that for a quadratic function $a + b^T x + \frac{1}{2} x^T A x$, the conjugate directions method minimizes the function in exactly n iterative steps. *Hint*:

$$\langle d^i, g^{i+1} \rangle = 0 = \langle d^i, Ax^{i+1} + b \rangle = \left\langle d^i, A\left(x^0 + \sum_{j=1}^{i} \alpha_j d^j\right) + b \right\rangle = 0, \text{ giving}$$

$$\alpha_i = -\frac{\langle d^i, Ax^0 + b \rangle}{\langle d^i, Ad^i \rangle}, \quad x^n = x^0 - \sum_{i=1}^{n} \frac{\langle d^i, Ax^0 + b \rangle}{\langle d^i, Ad^i \rangle} d^i.$$

15. Show that the minimum of $f(x)$ such that $\Phi(x) = 0$ satisfies $\frac{\partial}{\partial x} f(x) + \lambda \frac{\partial}{\partial x} \Phi(x) = 0$, where λ is called the *Lagrange multiplier*. Hence minimize $x^2 + y^2$ such that $x + 2y = 1$.

CHAPTER FIVE

MATRIX FUNCTIONS

5.1. Introduction

Matrix functions deserve a separate discussion, since they enter into many sorts of application. The reader is acquainted with some of the matrix functions like the inverse of a matrix, adjoint of a matrix, transpose of a matrix, etc. . . . In this chapter we generalize to consider many other functions. First we recall the most elementary ones, defined exactly like functions of a scalar, such as

$$e^A = I + A + \frac{A^2}{2!} + \cdots$$

$$\sin A = A - \frac{A^3}{3!} + \frac{A^5}{5!} - \cdots$$

$$\cos A = I - \frac{A^2}{2!} + \frac{A^4}{4!} - \cdots$$

$$\sinh A = A + \frac{A^3}{3!} + \cdots$$

$$\cosh A = I + \frac{A^2}{2!} + \cdots$$

$$\tan A = (\cos A)^{-1} \sin A$$

$$\tanh A = (\cosh A)^{-1} \sinh A$$

$$\log (I + A) = A - \frac{A^2}{2} + \frac{A^3}{3} - \cdots$$

The problem of defining matrix functions of general matrices is rather more difficult than it might seem at first glance. For example with functions expanded by Taylor's series, the problem of convergence should be studied. A simple example of that is the function

$$(I - A)^{-1} = I + A + A^2 + \cdots$$

whose expansion is only valid under conditions placed on A. And even if the convergence criterion is satisfied, it is still a problem to compute some matrix functions; one cannot accept that to compute $\sin A$ one needs to sum up the infinite series $A - \frac{A^3}{3!} + \cdots$. Before we proceed further, let us first define the basic matrix functions.

5.2. Adjoint of a matrix

The *adjoint* of a square matrix A, denoted by A^a, is a square matrix whose element (i,j) is equal to C_{ji} in the matrix A.

Example

$$A = \begin{bmatrix} 1 & 2 & 0 \\ 2 & 3 & -1 \\ 4 & 1 & 2 \end{bmatrix}, \quad A^a = \begin{bmatrix} 7 & -4 & -2 \\ -8 & 2 & 1 \\ -10 & 7 & -1 \end{bmatrix}$$

The relation between the adjoint of any matrix A and the determinant of A is established by the following theorem

THEOREM. $AA^a = |A|I$

Proof. Let the element (i,j) in AA^a be b_{ij}; then

$$b_{ij} = \sum_k a_{ik}a_{kj}^a$$

$$= \sum_k a_{ik}C_{jk}$$

And from Property 8 of determinants (Section 2.1) we obtain

$$b_{ij} = |A|, \quad i = j$$

$$= 0 \quad i \neq j$$

Hence

$$AA^a = \text{diag}(|A|, |A|, \ldots, |A|) = |A| \cdot I$$

5.3. Inverse of a matrix

The matrix A has a unique inverse only if it is square and nonsingular and is denoted by A^{-1}, where

$$AA^{-1} = A^{-1}A = I$$

However A can also possess what is termed a generalized inverse, even when it is singular or rectangular. We first start by citing the different methods for calculating the unique inverse of a nonsingular matrix A.

First method. Using the following theorem

$$AA^a = |A| \cdot I$$

we obtain

$$A^{-1} = \frac{A^a}{|A|}$$

Example

$$A = \begin{bmatrix} 1 & 2 & 1 \\ 0 & 1 & -1 \\ 1 & 3 & 1 \end{bmatrix}, \quad A^{-1} = \begin{bmatrix} 4 & 1 & -3 \\ -1 & 0 & 1 \\ -1 & -1 & 1 \end{bmatrix}$$

Second method. If a nonsingular matrix is transformed into a unit matrix by a series of elementary row operations, the same series of elementary row operations operating on a unit matrix will give the inverse of A, i.e. if

$$R_1 R_2 \ldots R_q A = I$$

then

$$A^{-1} = R_1 R_2 \ldots R_q I$$

which means that we operate simultaneously on both A and I; by so doing A is transformed into a unit matrix and I will automatically be transformed into the inverse of A.

Example. Find the inverse of A, if

$$A = \begin{bmatrix} 1 & 2 & 1 \\ 0 & 1 & -1 \\ 1 & 3 & 1 \end{bmatrix}$$

We write

$$A = \begin{bmatrix} 1 & 2 & 1 \\ 0 & 1 & -1 \\ 1 & 3 & 1 \end{bmatrix}, \quad I = \begin{bmatrix} 1 & 0 & 0 \\ 0 & 1 & 0 \\ 0 & 0 & 1 \end{bmatrix}$$

First step:

$$\begin{bmatrix} 1 & 2 & 1 \\ 0 & 1 & -1 \\ 0 & 1 & 0 \end{bmatrix}, \quad \begin{bmatrix} 1 & 0 & 0 \\ 0 & 1 & 0 \\ -1 & 0 & 1 \end{bmatrix}$$

Second step:

$$\begin{bmatrix} 1 & 2 & 1 \\ 0 & 1 & -1 \\ 0 & 0 & 1 \end{bmatrix}, \begin{bmatrix} 1 & 0 & 0 \\ 0 & 1 & 0 \\ -1 & -1 & 1 \end{bmatrix}$$

Third step:

$$\begin{bmatrix} 1 & 2 & 1 \\ 0 & 1 & 0 \\ 0 & 0 & 1 \end{bmatrix}, \begin{bmatrix} 1 & 0 & 0 \\ -1 & 0 & 1 \\ -1 & -1 & 1 \end{bmatrix}$$

Fourth step:

$$\begin{bmatrix} 1 & 0 & 0 \\ 0 & 1 & 0 \\ 0 & 0 & 1 \end{bmatrix}, \begin{bmatrix} 4 & 1 & -3 \\ -1 & 0 & 1 \\ -1 & -1 & 1 \end{bmatrix}$$

Third method. By partitioning the matrix A into a finite set of rectangular blocks some of which are nonsingular, the inverse of A can be obtained by solving a few equations involving matrices of lower order than A.
Let

$$A = \begin{bmatrix} A_1 & A_2 \\ A_3 & A_4 \end{bmatrix}$$

and let

$$A^{-1} = B = \begin{bmatrix} B_1 & B_2 \\ B_3 & B_4 \end{bmatrix}$$

The problem becomes that of solving for B_1, B_2, B_3 and B_3 and B_4. We assume without loss of generality that A_1 and A_4 are nonsingular. If the latter are singular we can still find a nonsingular submatrix of A. After writing $AB = I$ the equations become:

$$A_1B_1 + A_2B_3 = I, \quad A_1B_2 + A_2B_4 = 0$$

$$A_3B_1 + A_4B_3 = 0, \quad A_3B_2 + A_4B_4 = I$$

Solving, we obtain

$$B_1 = (A_1 - A_2A_4^{-1}A_3)^{-1}$$

$$B_3 = -A_4^{-1}A_3B_1$$

$$B_4 = (A_4 - A_3A_1^{-1}A_2)^{-1}$$

$$B_2 = -A_1^{-1}A_2B_4$$

Example. Find the inverse of A, if

$$A = \begin{bmatrix} G & -\eta \\ \eta^T & R \end{bmatrix}$$

Applying the above formulae, we obtain

$$A^{-1} = \begin{bmatrix} B_1 & B_2 \\ B_3 & B_4 \end{bmatrix}$$

where

$$B_1 = (G + \eta R^{-1}\eta^T)^{-1}, \quad B_2 = G^{-1}\eta(R + \eta^T G^{-1}\eta)^{-1}$$
$$B_3 = -R^{-1}\eta^T(G + \eta R^{-1}\eta^T)^{-1}, \quad B_4 = (R + \eta^T G^{-1}\eta)^{-1}$$

The above three methods are the most used ones; however, the first is a very tedious exercise especially for higher-order matrices and is only preferred when the matrix is of order less than five.

Before we move to dişcuss the generalized inverse of any matrix A, we should only mention that due to round-off errors, the inverse obtained using the above methods is only approximate. If we desire to improve its accuracy, then we could use a method of successive approximation. For example we can define an error matrix F_0 such that

$$F_0 = I - AB_0$$

The elements of F_0 are small, but they are zero only if B_0 is the correct inverse of A. Constructing the two sequences

$$\begin{aligned} F_1 &= I - AB_1, & B_1 &= B_0 + B_0F_0 \\ F_2 &= I - AB_2, & B_2 &= B_1 + B_1F_1 \\ &\vdots & &\vdots \end{aligned}$$

we can easily show, through simple manipulations, that

$$B_1 = A^{-1}(I - F_0)(I + F_0) = A^{-1}(I - F_0^2)$$
$$B_2 = A^{-1}(I - F_0^2)(I + F_1) = A^{-1}(I - F_0^2)(2I - I + F_0^2) = A^{-1}(I - F_0^4)$$

In general

$$B_m = A^{-1}(I - F_0^{2m})$$

If m is taken large enough, B_m will then approximate through iteration to A^{-1}, since

$$\| F_0 \| < 1$$

For further reading about correcting matrix inverses, refer to Demidovich and Maron (1973), p. 316.

Usually engineers and scientists, when asked why they need to compute A^{-1}, answer that they want to compute $A^{-1}b$, where b is a vector. But this is impractical since to compute

$$x = A^{-1}b$$

is synonymous with solving the set of linear equations

$$Ax = b$$

and this can be done with much saving of computation. The inverse is usually needed when we require x for different vectors b. In this case, it would be worth while to compute A^{-1} and then to do the multiplication.

There are many methods for computing a matrix inverse; Householder (1975), Chapter 5, mentioned only a few. Methods relying upon the eigenvalue problem of the matrix are discussed in the next sections.

THEOREM. Any matrix A has a generalized inverse A^i, if and only if

$$AA^iA = A$$

A^i is then given by

$$A^i = P\begin{bmatrix} I_\rho & U \\ V & W \end{bmatrix} R$$

where ρ is the rank of A, and P and R are such that

$$RAP = \begin{bmatrix} I_\rho & 0 \\ 0 & 0 \end{bmatrix}$$

U, V and W are arbitrary matrices.

The proof of the theorem is straightforward, and is left as an exercise for the reader.

Example

$$A = \begin{bmatrix} 1 & 2 & 1 \\ 0 & 1 & -2 \end{bmatrix} \Rightarrow R = \begin{bmatrix} 1 & -2 \\ 0 & 1 \end{bmatrix}, \quad P = \begin{bmatrix} 1 & 0 & -5 \\ 0 & 1 & 2 \\ 0 & 0 & 1 \end{bmatrix}$$

Therefore

$$A^i = \begin{bmatrix} 1 & 0 & -5 \\ 0 & 1 & 2 \\ 0 & 0 & 1 \end{bmatrix}\begin{bmatrix} 1 & 0 \\ 0 & 1 \\ \nu_1 & \nu_2 \end{bmatrix}\begin{bmatrix} 1 & -2 \\ 0 & 1 \end{bmatrix} = \begin{bmatrix} 1-5\nu_1 & -2+10\nu_1-5\nu_2 \\ 2\nu_1 & -4\nu_1+2\nu_2+1 \\ \nu_1 & -2\nu_1+\nu_2 \end{bmatrix}$$

where ν_1 and ν_2 are arbitrary scalars.

The part of A^i which does not depend on U, V and W is of special importance. We will call it A^t. For the example above

$$A^t = \begin{bmatrix} 1 & -2 \\ 0 & 1 \\ 0 & 0 \end{bmatrix}$$

In general A^t is obtained from A^i by setting U, V and W to zero. Equivalently it is obtained by partitioning P and R in the form

$$P = [P_1 \,\vdots\, P_2], \quad R = \begin{bmatrix} R_1 \\ \hdashline R_2 \end{bmatrix}$$

where P_1 and R_1 are respectively of order (n, ρ) and (ρ, m), if A is of order (m, n). A^t is then given by

$$A^t = P_1 R_1$$

The generalized inverse A^i of a matrix A could be well used to solve a set of linear simultaneous equations

$$Ax = b$$

where A is of order (m, n). The solution is given by

$$x = A^i b.$$

Substituting for A^i, we obtain

$$x = A^t b + f(U, V, W)b$$

Obviously we should expect that

$$Af(U, V, W)b = 0$$

$$AA^t b = b$$

so that we come back to the form of the set of equations. For the first condition, $f(U, V, W)b$ can be represented in the form

$$f(U, V, W)b = (I - A^t A)z$$

where z is an arbitrary vector which accounts for U, V and W. The reason is that

$$AA^i A = AA^t A = A$$

i.e. that

$$A(I - A^t A) = 0$$

As for the second condition, it is actually satisfied if

$$\rho(A) = \rho(A \,\vdots\, b)$$

i.e. if the equations are consistent. To show that, we substitute for A and A^t in the condition to give

$$AA^t b = R^{-1}\begin{bmatrix} I_\rho & 0 \\ 0 & 0 \end{bmatrix} P^{-1}P\begin{bmatrix} I_\rho & 0 \\ 0 & 0 \end{bmatrix} Rb = b$$

i.e.

$$\begin{bmatrix} I_\rho & 0 \\ 0 & 0 \end{bmatrix} Rb = Rb$$

But this is true from consistency of the equations; for the matrix R, when reducing A into an echelon form, reduces as well the vector b to a vector having its last $m - \rho$ elements equal to zero, i.e.

$$Rb = \begin{bmatrix} b^1 \\ --- \\ 0 \end{bmatrix}$$

Therefore we conclude that the set of linear simultaneous equations has a general solution in the form

$$x = A^t b + (I - A^t A)z$$

Example. Solve the equations

$$\begin{bmatrix} 10 & 0 & 5 & 0 \\ 4 & -2 & 2 & 1 \\ 2 & 4 & 1 & -2 \end{bmatrix}\begin{bmatrix} x_1 \\ x_2 \\ x_3 \\ x_4 \end{bmatrix} = \begin{bmatrix} 1 \\ 2 \\ -3 \end{bmatrix}$$

We determine R and P so that RAP becomes the normal form of A, giving

$$R = \begin{bmatrix} 1/10 & 0 & 0 \\ 1/5 & -\tfrac{1}{2} & 0 \\ -1 & 2 & 1 \end{bmatrix}, \quad P = \begin{bmatrix} 1 & 0 & -\tfrac{1}{2} & 0 \\ 0 & 1 & 0 & \tfrac{1}{2} \\ 0 & 0 & 1 & 0 \\ 0 & 0 & 0 & 1 \end{bmatrix}$$

Hence

$$A^t = \begin{bmatrix} 1 & 0 \\ 0 & 1 \\ 0 & 0 \\ 0 & 0 \end{bmatrix}\begin{bmatrix} 1/10 & 0 & 0 \\ 1/5 & -\tfrac{1}{2} & 0 \end{bmatrix} = \begin{bmatrix} 1/10 & 0 & 0 \\ 1/5 & -\tfrac{1}{2} & 0 \\ 0 & 0 & 0 \\ 0 & 0 & 0 \end{bmatrix}$$

So we obtain

$$I - A^t A = \begin{bmatrix} 0 & 0 & -\tfrac{1}{2} & 0 \\ 0 & 0 & 0 & \tfrac{1}{2} \\ 0 & 0 & 1 & 0 \\ 0 & 0 & 0 & 1 \end{bmatrix}$$

and the general solution becomes

$$\begin{bmatrix} x_1 \\ x_2 \\ x_3 \\ x_4 \end{bmatrix} = \begin{bmatrix} 1/10 & 0 & 0 \\ 1/5 & -½ & 0 \\ 0 & 0 & 0 \\ 0 & 0 & 0 \end{bmatrix} \begin{bmatrix} 1 \\ 2 \\ -3 \end{bmatrix} + \begin{bmatrix} 0 & 0 & -½ & 0 \\ 0 & 0 & 0 & ½ \\ 0 & 0 & 1 & 0 \\ 0 & 0 & 0 & 1 \end{bmatrix} \begin{bmatrix} z_1 \\ z_2 \\ z_3 \\ z_4 \end{bmatrix}$$

$$= \begin{bmatrix} 1/10 \\ -4/5 \\ 0 \\ 0 \end{bmatrix} + z_3 \begin{bmatrix} -½ \\ 0 \\ 1 \\ 0 \end{bmatrix} + z_4 \begin{bmatrix} 0 \\ ½ \\ 0 \\ 1 \end{bmatrix}$$

which is the same solution obtained before by the Gauss elimination method, treating the same example of Section 3.9b and taking $z_3 = -c_1, z_4 = -c_2$.

The reader should notice that the form of the general solution for x, that is

$$x = A^t b + (I - A^t A)z$$

is made possible only if the equations are consistent. But interestingly enough for the case where A is of order (m, n) and $m > n$, (see Section 3.9c), the equations have no exact solution, since in general

$$\rho(A) \neq \rho(A \vdots b)$$

and only an approximate solution for x to minimize $\| Ax - b \|_2$ is found. The best solution for x, if A and b are real, is given by

$$A^T A x = A^T b,$$

where the matrix $A^T A$ is nonsingular if $\rho(A) = n$. If $\rho(A) < n, A^T A$ is singular; however the above equations still have a solution for all vectors b, since x in this case is no more than $A^i b$, for

$$A^T A A^i b = A^T b$$

even if the original equations are inconsistent. The reason is that

$$A^T (I - AA^i) = 0$$

even if

$$AA^i b \neq b.$$

To show that

$$A^T (I - AA^i) = 0$$

it is not sufficient to rely on the definition of generalized inverse mentioned in the last theorem, since it only provides us with the equality

$$AA^i A = A$$

And if we take the transpose of the above equality, we will not arrive at proving that

$$A^T A A^i = A^T$$

unless

$$AA^i = (AA^i)^T.$$

But this is generally possible if one defines a generalized inverse which satisfies $AA^iA = A$ and $AA^i = (AA^i)^*$.

If a generalized inverse is found to satisfy not only $AA^iA = A$ and $AA^i = (AA^i)^*$ but also $A^iAA^i = A^i$ and $A^iA = (A^iA)^*$, then such a generalized inverse is uniquely determined. It is called the *pseudo-inverse* (*Penrose–Moore inverse*) of A denoted by $A\dagger$. It is therefore obvious that to find the pseudo-inverse of a matrix, one should apply more conditions on A^i defined in the last theorem. This is possible, since the matrices R and P are not unique and U, V and W are arbitrary. So if A^i is defined as

$$A^i = P\begin{bmatrix} I_\rho & U \\ V & W \end{bmatrix}R, \quad RAP = \begin{bmatrix} I_\rho & 0 \\ 0 & 0 \end{bmatrix}$$

then applying also $A^iAA^i = A^i$ leads us to

$$W = VU.$$

A generalized inverse which satisfies $AA^iA = A$ and $A^iAA^i = A^i$ is called a *reflexive generalized* inverse of A. Finally if we apply $AA^i = (AA^i)^*$ and $A^iA = (A^iA)^*$ we are led to conditions on U, V, R and P. The reader can check that these last conditions imply that

$$R^{-1}\begin{bmatrix} I & U \\ 0 & 0 \end{bmatrix}R \text{ and } P\begin{bmatrix} I & 0 \\ V & 0 \end{bmatrix}P^{-1}$$

are both Hermitian. Indeed they are; if $U = 0$, $V = 0$, R is decomposed into the product of a diagonal matrix and a unitary matrix and P is decomposed into the product of a unitary matrix and a diagonal matrix (see Exercise 4.10.38).

For computational aspects of the Penrose–Moore inverse, the reader is referred to Stoer and Bulirsch (1980), pp. 330–332, and Urquehart (1968). Pease (1965), p. 282, obtained the Penrose–Moore inverse using the minimal polynomial.

As for the solution of the set of linear simultaneous equations, it still takes the form

$$x = A\dagger b + (I - A\dagger A)z$$

Now we treat the same problem as above, using the pseudo-inverse of A.

Example. Solve the equations

$$\begin{bmatrix} 10 & 0 & 5 & 0 \\ 4 & -2 & 2 & 1 \\ 2 & 4 & 1 & -2 \end{bmatrix}\begin{bmatrix} x_1 \\ x_2 \\ x_3 \\ x_4 \end{bmatrix} = \begin{bmatrix} 1 \\ 2 \\ -3 \end{bmatrix}$$

We first obtain A^*A and AA^* to be

$$A^*A = \begin{bmatrix} 120 & 0 & 60 & 0 \\ 0 & 20 & 0 & -10 \\ 60 & 0 & 30 & 0 \\ 0 & -10 & 0 & 5 \end{bmatrix}, \quad AA^* = \begin{bmatrix} 125 & 50 & 25 \\ 50 & 25 & 0 \\ 25 & 0 & 25 \end{bmatrix}$$

whose eigenvalues and eigenvectors are respectively

$$\begin{aligned} \lambda_1 &= 150 \\ \lambda_2 &= 25 \\ \lambda_3 &= 0 \\ \lambda_4 &= 0 \end{aligned} \quad u^1 = \begin{bmatrix} 2 \\ 0 \\ 1 \\ 0 \end{bmatrix}, \quad u^2 = \begin{bmatrix} 0 \\ 2 \\ 0 \\ -1 \end{bmatrix} \quad u^3 = \begin{bmatrix} 1 \\ 0 \\ -2 \\ 0 \end{bmatrix},$$

$$u^4 = \begin{bmatrix} 0 \\ 1 \\ 0 \\ 2 \end{bmatrix}$$

and

$$\begin{aligned} \lambda_1 &= 150 \\ \lambda_2 &= 25 \\ \lambda_3 &= 0 \end{aligned} \quad u^1 = \begin{bmatrix} 5 \\ 2 \\ 1 \end{bmatrix}, \quad u^2 = \begin{bmatrix} 0 \\ -1 \\ 2 \end{bmatrix}, \quad u^3 = \begin{bmatrix} 1 \\ -2 \\ -1 \end{bmatrix}$$

The matrices R and P can be taken as

$$R = \begin{bmatrix} \frac{5}{150} & \frac{2}{150} & \frac{1}{150} \\ 0 & -\frac{1}{25} & \frac{2}{25} \\ 1 & -2 & -1 \end{bmatrix}, \quad P = \begin{bmatrix} 2 & 0 & 1 & 0 \\ 0 & 2 & 0 & 1 \\ 1 & 0 & -2 & 0 \\ 0 & -1 & 0 & 2 \end{bmatrix}$$

Therefore

$$A^t = \begin{bmatrix} 2 & 0 \\ 0 & 2 \\ 1 & 0 \\ 0 & -1 \end{bmatrix} \begin{bmatrix} \frac{5}{150} & \frac{2}{150} & \frac{1}{150} \\ 0 & \frac{-1}{25} & \frac{2}{25} \end{bmatrix} = \begin{bmatrix} \frac{10}{150} & \frac{4}{150} & \frac{2}{150} \\ 0 & -\frac{2}{25} & \frac{4}{25} \\ \frac{5}{150} & \frac{2}{150} & \frac{1}{150} \\ 0 & \frac{1}{25} & \frac{-2}{25} \end{bmatrix}$$

and

$$I - A\dagger A = \begin{bmatrix} \frac{1}{5} & 0 & \frac{-2}{5} & 0 \\ 0 & \frac{1}{5} & 0 & \frac{2}{5} \\ \frac{-2}{5} & 0 & \frac{4}{5} & 0 \\ 0 & \frac{2}{5} & 0 & \frac{4}{5} \end{bmatrix}$$

The general solution becomes

$$\begin{bmatrix} x_1 \\ x_2 \\ x_3 \\ x_4 \end{bmatrix} = \begin{bmatrix} 10/150 & 4/150 & 2/150 \\ 0 & -2/25 & 4/25 \\ 5/150 & 2/150 & 1/150 \\ 0 & 1/25 & -2/25 \end{bmatrix} \begin{bmatrix} 1 \\ 2 \\ -3 \end{bmatrix} +$$

$$+ \begin{bmatrix} 1/5 & 0 & -2/5 & 0 \\ 0 & 1/5 & 0 & 2/5 \\ -2/5 & 0 & 4/5 & 0 \\ 0 & 2/5 & 0 & 4/5 \end{bmatrix} \begin{bmatrix} z_1 \\ z_2 \\ z_3 \\ z_4 \end{bmatrix}$$

$$= \begin{bmatrix} 12/150 \\ -16/25 \\ 6/150 \\ 8/25 \end{bmatrix} + z_1 \begin{bmatrix} 1/5 \\ 0 \\ -2/5 \\ 0 \end{bmatrix} + z_2 \begin{bmatrix} 0 \\ 1/5 \\ 0 \\ 2/5 \end{bmatrix} + z_3 \begin{bmatrix} -2/5 \\ 0 \\ 4/5 \\ 0 \end{bmatrix} +$$

$$+ z_4 \begin{bmatrix} 0 \\ 2/5 \\ 0 \\ 4/5 \end{bmatrix}$$

The reader should not get confused by the number of solutions of the homogeneous system of equations, since the four last vectors above have rank two, meaning that we have only, as we expected, two independent solutions. Note also that the form of solution obtained above does not differ from that obtained before, treating the same problem. This is made possible by taking for example

$$z_1 = \frac{1}{10} - \frac{5}{2}\alpha$$

$$z_2 = \frac{-4}{5} + \frac{5}{2}\beta$$

$$z_3 = 0$$

$$z_4 = 0$$

5.4. Resolvent of a matrix, $(\lambda I - A)^{-1}$

The resolvent of a matrix comes into many sorts of application. In this section we mention two applications, one related to a proof of the Cayley–Hamilton theorem as an alternative to the previous proof, and the other related to computing the minimal polynomial of a matrix by a method much simpler than checking the function $\prod_{i=1}^{r}(A - \lambda_i I)$, which was explained before.

The resolvent of a matrix A can be computed using the method of the adjoint as follows:

$$(\lambda I - A)^{-1} = \frac{(\lambda I - A)^a}{|\lambda I - A|}$$

$$= \frac{\lambda^{n-1}B_{n-1} + \lambda^{n-2}B_{n-2} + \cdots + \lambda B_1 + B_0}{\lambda^n + a_{n-1}\lambda^{n-1} + \cdots + a_1\lambda + a_0}$$

where the polynomial in the denominator is the characteristic polynomial of A. To compute $a_{n-1}, \ldots, a_0$ and $B_{n-1}, B_{n-2}, \ldots, B_0$ we use Leverrier's algorithm, after equating powers of λ to obtain:

$$B_{n-1} = I$$
$$B_{n-2} = a_{n-1}I + AB_{n-1}$$
$$B_{n-3} = a_{n-2}I + AB_{n-2}$$
$$\vdots$$
$$B_1 = a_2 I + AB_2$$
$$B_0 = a_1 I + AB_1$$
$$0 = a_0 I + AB_0$$

where $a_{n-1}, a_{n-2}, \ldots, a_0$ can be looked up in Section 4.2. What is interesting is that from the above equation we can arrive very simply at a proof of the Cayley–Hamilton theorem. Starting from the last equation and moving upward by substituting for B_0 and then for B_1 etc. . . . , we obtain:

$$0 = a_0 I + A(a_1 I + AB_1)$$
$$= a_0 I + a_1 A + A^2(a_2 I + AB_2)$$
$$\vdots$$
$$= a_0 I + a_1 A + a_2 A^2 + \cdots + a_{n-1}A^{n-1} + A^n$$

which is the statement of the Cayley–Hamilton theorem. Another application of the resolvent is finding the minimal polynomial of A. The latter is obtained by considering the quotient

$$\frac{(\lambda I - A)^a}{|\lambda I - A|}$$

and upon cancelling common factors from both the numerator and the denominator, what remains in the denominator is the minimal polynomial of A. For example, to determine the minimal polynomial of the matrix

$$A = \begin{bmatrix} 2 & 1 & 0 & 0 \\ 0 & 2 & 1 & 0 \\ 0 & 0 & 2 & 0 \\ 0 & 0 & 0 & 2 \end{bmatrix}$$

we consider

$$\frac{(\lambda I - A)^a}{|\lambda I - A|} = \frac{\begin{bmatrix} (\lambda-2)^3 & (\lambda-2)^2 & (\lambda-2) & 0 \\ 0 & (\lambda-2)^3 & (\lambda-2)^2 & 0 \\ 0 & 0 & (\lambda-2)^3 & 0 \\ 0 & 0 & 0 & (\lambda-2)^3 \end{bmatrix}}{(\lambda-2)^4}$$

$$= \frac{\begin{bmatrix} (\lambda-2)^2 & (\lambda-2) & 1 & 0 \\ 0 & (\lambda-2)^2 & (\lambda-2) & 0 \\ 0 & 0 & (\lambda-2)^2 & 0 \\ 0 & 0 & 0 & (\lambda-2)^2 \end{bmatrix}}{(\lambda-2)^3}$$

Hence $(\lambda - 2)^3$ is the minimal polynomial.
To prove the above proposition, write

$$(\lambda I - A)(\lambda I - A)^a = |\lambda I - A| \cdot I = \prod_{i=1}^{n} (\lambda - \lambda_i) \cdot I$$

So if $(\lambda I - A)^a$ has common factors with $|\lambda I - A|$, by cancelling them from both sides we obtain

$$(\lambda I - A)(\lambda I - A)'^a = \prod_{i=1}^{r} (\lambda - \lambda_i) \cdot I$$

The reader should try to prove that $(\lambda_i I - A)$ is proportional, in the simple case, to

$$\frac{d^{r-2}}{d\lambda^{r-2}} (\lambda I - A)'^a |_{\lambda=\lambda_i}$$

Differentiating the above equation $(r - 1)$ times w.r.t. to λ gives

$$\prod_{i=1}^{r} (\lambda_i I - A) = 0$$

meaning that r is the order of the minimal polynomial.

One relevant application of the resolvent to the theory of linear dynamical systems is that it appears in the analysis of such systems when using Laplace transform techniques; see Zadeh and Desoer (1963), p. 594. By cancelling common factors between the numerator and the denominator of $(\lambda I - A)^{-1}$ we conclude that it is the minimal polynomial and not the characteristic polynomial which characterizes the dynamics of a system.

Another beautiful application of the resolvent is the calculation of A^{-1}. It is deduced from the relation

$$(\lambda I - A)^{-1} |_{\lambda=0} = \frac{B_0}{a_0}$$

and computed to great accuracy. The only drawback of this method is its slowness.

Exercises 5.4.

1. Find $|A^a|$.
2. Find $(A^a)^a$.
3. Show that $(AB)^a = B^a A^a$.
4. Show that $(A^m)^a = (A^a)^m$.
5. Show that $(A^a)^* = (A^*)^a$.
6. Show that the adjoint of a Hermitian matrix is Hermitian.
7. Find the inverse of A and B, if

$$A = \begin{bmatrix} 1 & 0 & -1 \\ 2 & 0 & 3 \\ -1 & 1 & 0 \end{bmatrix}, \quad B = \begin{bmatrix} 1 & 2 & -2 \\ 2 & -1 & 1 \\ 0 & -1 & -5 \end{bmatrix}$$

Can the inverse of B be obtained easily making use of the orthogonality of its column vectors?

8. If $A = \text{quasidiag}(A_1, A_2, \ldots, A_m)$, show that $A^{-1} = \text{quasidiag}(A_1^{-1}, A_2^{-1}, \ldots, A_m^{-1})$.
9. Show that $(AB)^{-1} = B^{-1}A^{-1}$.
10. Show that $(A^{-1})^* = (A^*)^{-1}$.
11. Show that $(A^a)^{-1} = (A^{-1})^a$.
12. If H is Hermitian, show that H^{-1} is also Hermitian.
13. If H is positive definite, show that H^{-1} is also positive definite.
14. Show that the inverse of an elementary operation is an elementary operation.
15. If A is nonsingular, show that $\begin{bmatrix} A & B \\ C & D \end{bmatrix}$ is also nonsingular if $|D - CA^{-1}B| \neq 0$.
16. Obtain a relation between $\lambda(A^a)$ and $\lambda(A)$.
17. Explain how to use the power method to compute the smallest eigenvalue in magnitude $\check{\lambda}(A)$. *Hint*: compute $\hat{\lambda}(A^{-1})$.
18. If S is real skew-symmetric, show that $(I - S)(I + S)^{-1}$ is orthogonal.
19. Obtain a reflexive generalized inverse of $A = \begin{bmatrix} 1 & -1 \\ 0 & 1 \\ 1 & 2 \end{bmatrix}$.
20. Show that if $A = \text{diag}(\alpha_1, \ldots, \alpha_r, 0, \ldots, 0)$, then $A^\dagger = \text{diag}(1/\alpha_1, 1/\alpha_2, \ldots, 1/\alpha_r, 0, \ldots, 0)$.

21. Show that $(A\dagger)^* = (A^*)\dagger$.
22. Show that the pseudo-inverse of a Hermitian matrix is Hermitian.
23. Is $(AB)\dagger = B\dagger A\dagger$?
24. If A is nonsingular, show that $A\dagger = A^{-1}$.
25. If $A = \text{quasidiag}(A_1, A_2, \ldots, A_m)$ show that $A\dagger = \text{quasidiag}(A_1^\dagger, A_2^\dagger, \ldots, A_m^\dagger)$.
26. If A is any complex matrix of order (m, n) and $U^*AV = Z = \begin{bmatrix} D & 0 \\ 0 & 0 \end{bmatrix}$, where

U and V are unitary and whose columns are respectively the eigenvectors of AA^* and A^*A, $D = \text{diag}(\sigma_1, \ldots, \sigma_r)$ where $\sigma_1, \ldots, \sigma_r > 0$ being the non-vanishing singular values of A and r is the rank of A. Show that $A\dagger = VZ\dagger U^*$, where

$Z\dagger = \begin{bmatrix} D^{-1} & 0 \\ 0 & 0 \end{bmatrix}$ and is of dimension (n, m). This describes a systematic computation scheme of $A\dagger$.

27. If A is singular, show that there exist singular matrices H and U, such that $A = UH$, where H is Hermitian non-negative definite such that $H^2 = A^*A$ and $U = AH\dagger$, with $U^* = U\dagger$.
28. A normal matrix A is called positive semidefinite if Re $x^*Ax \geqslant 0, \forall x \neq 0$. Show that A is positive semidefinite if and only if either of the following conditions is satisfied.

(i) Re $\lambda \geqslant 0$.

(ii) A can be put in the form $A = QD_uQ^*$, where Q is a nonsingular matrix, such that Q^*Q is diagonal; and $D_u = \text{diag}(e^{i\theta_1}, e^{i\theta_2}, \ldots, e^{i\theta_r}, 0, \ldots, 0)$, $i = \sqrt{-1}$, $-\dfrac{\pi}{2} < \theta_k < \dfrac{\pi}{2}$, where $e^{i\theta_k}$ is the eigenvalue of a singular matrix U, such that $A = HU = UH$, where H is the Hermitian non-negative definite square root of A^*A and $U = H\dagger A = AH\dagger$, with $U^* = U\dagger$, and r is the rank of A.

29. If $\rho(A) = r < n$ and supposing that we know the $n \times r$ matrix B and $r \times n$ matrix C both of rank r such that $A = BC$, show that $A\dagger = C^*(CC^*)^{-1}(B^*B)^{-1}B^*$.
30. Find the condition for the system of linear simultaneous equations

$$\begin{bmatrix} A & B \\ C & D \end{bmatrix}\begin{bmatrix} x \\ y \end{bmatrix} = \begin{bmatrix} u \\ v \end{bmatrix}$$

to have a unique solution for x and more than one solution for y.

5.5. Sequence and series of matrices

Let some process of construction yield a succession of matrices in the form

$$A_1, A_2, A_3, \ldots, A_n, \ldots$$

where it is assumed that every matrix A_i is followed by other terms. A succession of matrices formed in this way is called a *sequence*, and is denoted by $\{A_i\}$. An example of a sequence is

$$\begin{bmatrix} 1 & 1 \\ 1/2 & 0 \end{bmatrix}, \begin{bmatrix} 2 & -2 \\ 1/3 & 1 \end{bmatrix}, \begin{bmatrix} 3 & 3 \\ 1/4 & 0 \end{bmatrix}, \begin{bmatrix} 4 & -4 \\ 1/5 & 1 \end{bmatrix}, \ldots$$

It may happen that one can find a positive number m such that

$$\| A_n \| < m, \quad \forall n.$$

In that case the sequence $\{A_i\}$ is said to be bounded. For example the sequence

$$\begin{bmatrix} -1 & 2 \\ 0 & 1/2 \end{bmatrix}, \begin{bmatrix} 1 & 2 \\ 0 & 2/3 \end{bmatrix}, \begin{bmatrix} -1 & 2 \\ 0 & 3/4 \end{bmatrix}, \ldots, \begin{bmatrix} (-)^n & 2 \\ 0 & \dfrac{n}{n+1} \end{bmatrix}, \ldots$$

is bounded, since

$$\| A_n \|_1 = 2 + \frac{n}{n+1} < 3$$

DEFINITION 1. A sequence $\{A_i\}$ is called a *null sequence*, if subsequent to the choice of a positive number ϵ, however small, one can find a positive number p, such that $\| A_n \| < \epsilon$ for all values of $n \geqslant p$. An example of a null sequence is the sequence

$$\begin{bmatrix} 1 & 0 \\ 0 & 1 \end{bmatrix}, \begin{bmatrix} 1/2 & 0 \\ 0 & 1/4 \end{bmatrix}, \begin{bmatrix} 1/3 & 0 \\ 0 & 1/9 \end{bmatrix}, \begin{bmatrix} 1/4 & 0 \\ 0 & 1/16 \end{bmatrix}, \ldots$$

since one can always find a positive number ϵ such that $\| A_n \| < \epsilon$. The reader is asked to find the relation between n and ϵ.

DEFINITION 2. If $\{A_i\}$ is a given sequence, and there exists a matrix L such that the sequence $\{A_i - L\}$ is a null sequence, $\{A_i\}$ is said to be convergent, and the matrices A_i of the sequence are said to approach the limit L.

Recalling the definition of a null sequence, it is clear that if the limit of the sequence $\{A_i\}$ is L, then corresponding to any positive integer ϵ, a positive p can be assigned such that

$$\| L - A_n \| < \epsilon, \quad \forall n \geqslant p.$$

The reader may prove that the above statement is synonymous with

$$\lim_{n \to \infty} A_n = L$$

It is clear that a null sequence is a convergent sequence with the value zero as the limit of the sequence. Moreover, every convergent sequence is bounded. The reader should have no difficulty in establishing this property, but is the opposite true? *Hint*: think of an example in which A_i is bounded but not convergent.

DEFINITION 3. Let $A_1, A_2, \ldots, A_n, \ldots$ be any sequence of matrices; then the sum

$$\sum_{i=1}^{\infty} A_i = A_1 + A_2 + \cdots$$

is called an *infinite series.*

The similarity between series and sequences can be seen if we form a new sequence $\{S_i\}$ whose elements are sums of a finite number of the terms of the series

$$\sum_{i=1}^{\infty} A_i,$$

i.e.

$$S_1 = A_1$$
$$S_2 = A_1 + A_2$$
$$S_3 = A_1 + A_2 + A_3$$
$$\vdots$$
$$S_n = A_1 + A_2 + \cdots + A_n$$
$$\vdots$$

The matrices S_i will be called the ith *partial sums* of the series $\sum_{i=1}^{\infty} A_i$.

Now, we are in a position to define the convergence of the series as the convergence of its partial sums S_i. The series is therefore said to be convergent if the sequence of partial sums $\{S_i\}$ is convergent, i.e.

$$\| S - S_n \| < \epsilon, \quad \forall n \geqslant p.$$

The matrix S is called the *sum* of the series, and it is obviously the limit of its partial sums S_n.

Although the series can be treated as a sequence of its partial sums S_i for proving its convergence yet one should rely on other tests of convergence. The reason is that it is generally difficult to find S_i except for rare cases.

These are many tests of convergence, but we will recall only the most popular one, which is the d'Alembert test; it states that if

$$\frac{\| A_{n+1} \|}{\| A_n \|} < 1, \quad \forall n$$

then the series $\sum_{i=1}^{\infty} A_n$ is absolutely convergent. The reader is asked to investigate the case when the above quotient is greater than or equal to one. The proof of d'Alembert's test is easy and relies upon comparing the series of the norms with the geometric series, and it is left as an exercise for the reader.

Example. Test the convergence of the series

$$(I - A)^{-1} = I + A + A^2 + \cdots$$

The series is convergent if the series

$$1 + \| A \| + \| A \|^2 + \cdots \text{ is convergent}$$

Therefore the series is convergent if

$$\| A \| < 1$$

Note that this condition of convergence is severe and a better condition is given by the following theorem.

THEOREM. If the series $f(\lambda) = \sum_{i=0}^{\infty} a_i\lambda^i$ is convergent for all values of λ such that $|\lambda| < R$, the matrix series $F(A) = \sum_{i=0}^{\infty} a_i A^i$ is also convergent if A has λ as an eigenvalue.

Proof. The proof follows directly by writing

$$A = TJT^{-1}$$

Therefore

$$F(A) = \sum_{i=0}^{\infty} a_i A^i = T \left[\begin{array}{cccc:ccc} f(\lambda) & f'(\lambda) & \cdots & \dfrac{f^{(m-1)}(\lambda)}{(m-1)!} & & & \\ & f(\lambda) & f'(\lambda) & \vdots & & 0 & \\ & & \ddots & f'(\lambda) & & & \\ & & & f(\lambda) & & & \\ \hdashline & & & & f(\lambda_{m+1}) & & \\ & 0 & & & & \ddots & \\ & & & & & & f(\lambda_n) \end{array}\right] T^{-1}$$

where λ is a non-semi-simple eigenvalue of A, and $f'(\lambda)$ denotes differentiation w.r.t. λ. So if $f(\lambda_i)$ is convergent for all values of $|\lambda_k| < R$, then so is $f'(\lambda_i), \ldots, f^{(m-1)}(\lambda_i)$, as the reader can show; consequently $F(A)$ is convergent.

Going back to the previous example, the series

$$I + A + A^2 + \cdots$$

is convergent upon convergence of the series

$$1 + \lambda_i + \lambda_i^2 + \cdots, \quad i = 1, \ldots, n$$

i.e. for the matrix series to be convergent, it suffices to ensure that

$$s(A) < 1$$

which is less restrictive than $\| A \| < 1$, since $s(A) \leqslant \| A \|$.

The remainder of this section is devoted to the study of series whose terms are functions of a real variable t. A series of functions of t

$$A_1(t) + A_2(t) + \cdots + A_n(t) + \cdots$$

will be denoted by the symbol $\sum_{i=1}^{\infty} A_n(t)$; and it will be assumed that the functions $A_n(t)$ are defined in the interval (a, b). If the given series converges in the interval (a, b) then it is said to be convergent in (a, b). A series $\sum_{n=1}^{\infty} A_n(t)$, convergent in (a, b), defines a function of t in that interval, which will be denoted by $S(t)$. The sum $S_n(t)$ of the first n terms of the series is called the nth partial sum, and the statement that the series converges for a given value of t, say $t = t_1$, means that

$$\lim_{n \to \infty} S_n(t_1) = S(t_1)$$

It should be noted carefully that the definition of convergence just given is concerned with the convergence at a given point $t = t_1$ of the interval (a, b). But the series $\sum_{i=1}^{\infty} A_i(t)$ is called *uniformly convergent* in the interval (a, b) if for any $\epsilon > 0$, there exists a number p, independent of the value of t in (a, b), such that

$$\| S(t) - S_n(t) \| < \epsilon, \quad \forall n \geqslant p$$

which means that the concept of uniform convergence is connected inescapably with the interval of convergence. To explain the matter further, although convergent series at any point $t \in (a, b)$ and uniformly convergent series in the interval (a, b) are both convergent, they differ in the behaviour of their nth partial sums. For convergent series only at $t \in (a, b)$ we find that it is impossible for $S_n(t)$ to lie always close to $S(t)$ irrespective of the value of $t \in (a, b)$; whereas a uniformly convergent series has the property that its nth partial sum always differs from $S(t)$ by a small controllable amount which does not depend on $t \in (a, b)$. Of course we may find in some cases that the series may be non-uniformly convergent in $t \in (a, b)$, only numerically convergent, but uniformly convergent in an interval (t_1, b), where $a < t_1 < b$. To explain the matter more, consider the two examples of Sokolnikoff (1939), pp. 253, 254, of scalar matrices:

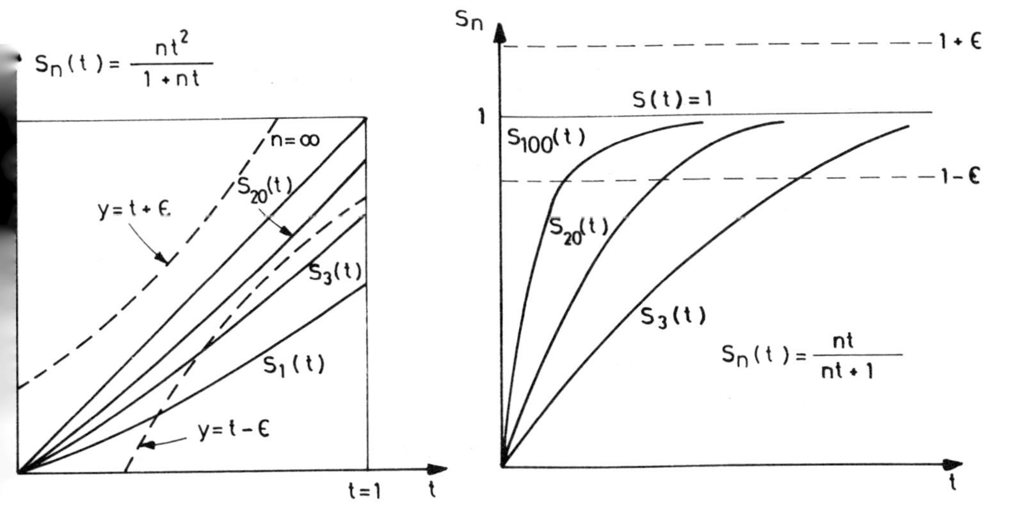

It is clear that the series $S_n(t) = nt/(nt + 1)$ is only numerically convergent for all $t \in (0, 1)$ but it is non-uniformly convergent since it is impossible to bound the nth partial sum between $S - \epsilon$ and $S + \epsilon$, whereas the series $S_n(t) = nt^2/(1 + nt)$ is uniformly convergent provided that n is taken large enough, which means that ϵ depends solely on n and not on t.

One last word should be said is that series which were found uniformly convergent in an interval (a, b) must necessarily be numerically convergent for all $t \in (a, b)$, as the reader can show; what remains now is to find conditions of uniform convergence of series that are better than just finding that ϵ is not a function of t.

One famous test of uniform convergence is that of Weierstrass, which asserts that the series

$$A_1(t) + A_2(t) + A_3(t) + \cdots$$

is uniformly and absolutely convergent in (a, b), if there exists a convergent series of positive constants

$$m_1 + m_2 + m_3 + \cdots$$

such that

$$\| A_i(t) \| < m_i$$

for all values of $t \in (a, b)$. To establish the theorem note that the series of ms is convergent, so that from Exercise 5.5.8 we obtain

$$m_{n+1} + m_{n+2} + \cdots + m_{n+k} < \epsilon$$

whenever $n \geqslant p$ and for all positive integers k. But each $A_i(t)$ is such that $\| A_i(t) \| < m_i$ for all values of $t \in (a, b)$; hence

$$\| A_{n+1}(t) \| + \| A_{n+2}(t) \| + \cdots + \| A_{n+k}(t) \| < \epsilon$$

and since the last sum is independent of t in the interval (a, b), the series $\sum_{i=1}^{\infty} A_i(t)$ is uniformly and absolutely convergent.

Example. Test for uniform convergence the series

$$\frac{A_1}{1^2} + \frac{A_2}{2^2} + \frac{A_3}{3^2} + \cdots$$

where

$$A_n = \begin{bmatrix} \sin nt & \cos nt \\ -\cos nt & \sin nt \end{bmatrix}$$

The matrix A_n is unitary; therefore $\| A_n(t) \| < m$ and the convergent series

$$\frac{m}{1^2} + \frac{m}{2^2} + \frac{m}{3^2} + \cdots$$

will serve as an m series for the given series. Therefore the given series converges uniformly and absolutely for any value of t.

One last remark about uniform convergence is that uniformly convergent series in an interval (a, b), when integrated term by term, converge to the integral of the function defined by the series.

Example. Test the convergence of the series e^{At}.

$$e^{At} = I + At + \frac{A^2 t^2}{2!} + \cdots$$

Applying the d'Alembert test on the series whose nth term is $\frac{\| A \|^n t^n}{n!}$ we find that

$$\frac{\| A \|^{n+1} t^{n+1}}{(n+1)!} \cdot \frac{n!}{\| A \|^n t^n} = \frac{\| A \| t}{(n+1)}$$

i.e.

$$\lim_{n \to \infty} \frac{\| A \| t}{(n+1)} = 0 < 1, \quad \forall t$$

which makes the series absolutely convergent for all values of t. To prove the uniform convergence of the series, we choose any interval for t, say $t \in [0, t_1]$. The nth term

$$\left\| \frac{A^n t^n}{n!} \right\| \leqslant \frac{\| A \|^n}{n!} t_1^n$$

Applying the Weierstrass test of uniform convergence on the series whose nth term is bounded by $\frac{\| A \|^n}{n!} t_1^n$, we conclude that e^{At} is uniformly convergent for all $t \in [0, t_1]$, and by increasing t_1 we realize that the series e^{At} is uniformly and absolutely convergent for all values of t.

Exercises 5.5.

1. If $\{A_i\}$ is a null sequence, show that $\lim_{n\to\infty} \| A_n \| = 0$ implies that $\lim_{n\to\infty} A_n = 0$.
2. If the terms of any null sequence $\{A_i\}$ are multiplied by a sequence of bounded factors $B_1, B_2, \ldots$, show that the resulting sequence $\{A_i B_i\}$ is a null sequence.
3. Show that a necessary and sufficient condition for the convergence of the sequence $\{A_i\}$ is that, for any $\epsilon > 0$, one can find a positive integer p such that $\| A_{n+k} - A_n \| < \epsilon$, $\forall n \geqslant p$, and for every positive integer k. *Hint*: write $A_{n+k} - A_n = (A_{n+k} - L) + (L - A_n)$.
4. A monotonic decreasing sequence $\{A_i\}$ is defined such that $A_i > A_{i+1}$. Show that if the sequence $\{A_i\}$ is monotonic decreasing and bounded below, i.e. $A_i > M$, it is convergent.
5. Show that the sequence $\{\sin nI\}$ is bounded but not convergent.
7. Show that a series of positive terms is necessarily convergent if its partial sums are bounded above. *Hint*: S_i is monotonic increasing; use the opposite of Exercise 4.
8. Show that a necessary and sufficient condition for the convergence of $\sum_{i=1}^{\infty} A_i$ is that, for any positive number ϵ, one can find a positive integer p, such that $\| S_{n+k} - S_n \| < \epsilon$. $\forall n \geqslant p$, and for every positive integer k.
9. Show that a necessary condition for convergence of $\sum_{n=1}^{\infty} A_n$ is that $\lim_{n\to\infty} A_n = 0$. *Hint*: use Exercise 8, take $k = 1$.
10. Show that if $\sum_{i=1}^{\infty} A_i$ is convergent, the sequence $\{S_i\}$ is bounded.
11. If the series $\sum_{i=1}^{\infty} A_i$ is such that $\sqrt[n]{\| A_n \|} = r < 1$, show that the series is absolutely convergent.
12. Obtain the nth partial sum of the series $\sum_{k=1}^{\infty} \sin kA$. *Hint*: obtain the imaginary part for $\sum_{k=1}^{n} e^{ikA}$, $i = \sqrt{-1}$.
13. Test for uniform convergence the series $\cosh At$.
14. Prove *Abel's test of uniform convergence*: the series $\sum_{n=1}^{\infty} A_n(t)$ converges uniformly in the interval $a \leqslant t \leqslant b$ if the functions $A_n(t)$ are of the form

$A_n(t) = B_n F_n(t)$, where $\sum_{n=1}^{\infty} B_n$ is convergent and the functions $F_n(t)$ are positive and $F_{n+1}(t) \leqslant F_n(t) < M$.

15. If A is nilpotent, show that $\sum_{n=0}^{\infty} a_n A^n$ converges for all bounded values a_n.

5.6. Computing matrix functions

Once a matrix function is found to exist it remains to derive suitable methods for its calculation. These are situations where the function is given by an infinite series which cannot be obviously summed. In this section we talk of two main methods for calculating different matrix functions. The first method relies upon the definition of the minimal polynomial of a matrix, while the second uses the spectral representation. We shall not of course derive all matrix functions, rather we will apply the techniques on some important examples like $A^{-1}, e^A, \sqrt{A}$. The reader can generalize the techniques explained on other functions of interest.

To compute A^{-1}, we can make use of the minimal polynomial as follows. If the minimal polynomial of A is of order m, with $m \geqslant n$, where n is the order of A, then

$$a_m A^m + a_{m-1} A^{m-1} + \cdots + a_1 A + a_0 = 0.$$

Multiplying both sides by A^{-1}, we obtain

$$A^{-1} = -\frac{1}{a_0}(a_1 I + a_2 A + \cdots + a_m A^{m-1})$$

with $a_0 \neq 0$, which means that, to calculate A^{-1}, we need to know the characteristic polynomial or the minimal polynomial as well as high powers of A until $n-1$ at most. It seems that the previous methods for calculating A^{-1} are much simpler, but this is not true if we have already solved the eigenvalue problem of A.

Example. Let

$$A = \begin{bmatrix} 5 & -2 & -4 \\ -2 & 2 & 2 \\ -4 & 2 & 5 \end{bmatrix} \quad \text{with } \lambda_1 = 10, \lambda_2 = \lambda_3 = 1$$

The matrix A is symmetric, hence semi-simple, and the minimal polynomial is $(\lambda - 10)(\lambda - 1) = \lambda^2 - 11\lambda + 10$.

$$A^{-1} = \frac{1}{10}(11I - A) = \frac{1}{10}\begin{bmatrix} 6 & 2 & 4 \\ 2 & 9 & -2 \\ 4 & -2 & 6 \end{bmatrix}$$

To calculate e^A, where A is of order n, we write the series

$$e^A = I + A + \frac{A^2}{2!} + \cdots$$

which we would not dare to sum. Before using our technique we distinguish between three cases. The first case is when A has distinct eigenvalues, and then the minimal polynomial is equal to the characteristic polynomial. In this case, from the Cayley–Hamilton theorem, the term A^n can be written in terms of $A, A^2, \ldots, A^{n-1}$, which gives

$$e^A = c_0 I + c_1 A + \cdots + c_{n-1} A^{n-1}$$

Post-multiplying both sides with u^i gives

$$e^{\lambda_i} u^i = (c_0 + c_1 \lambda_i + \cdots + c_{n-1} \lambda_i^{n-1}) u^i$$

and

$$e^{\lambda_i} = c_0 + c_1 \lambda_i + \cdots + c_{n-1} \lambda_i^{n-1}$$

Hence for all $\lambda_i = \lambda_1, \lambda_2, \ldots, \lambda_n$ we obtain

$$\begin{bmatrix} 1 & \lambda_1 & \lambda_1^2 & \cdots & \lambda_1^{n-1} \\ 1 & \lambda_2 & & \cdots & \\ \vdots & & & & \\ 1 & \lambda_n & & \cdots & \lambda_n^{n-1} \end{bmatrix} \begin{bmatrix} c_0 \\ c_1 \\ \vdots \\ c_{n-1} \end{bmatrix} = \begin{bmatrix} e^{\lambda_1} \\ e^{\lambda_2} \\ \vdots \\ e^{\lambda_n} \end{bmatrix}$$

The matrix on the left-hand side is a Vandermonde matrix, which we have met before, and is nonsingular for $\lambda_1 \neq \lambda_2 \neq \cdots \neq \lambda_n$; hence the coefficients $c_0, c_1, \ldots, c_{n-1}$ can be calculated uniquely.

Example. Calculate e^A if $A = \begin{bmatrix} 1 & 2 \\ 2 & -1 \end{bmatrix}$

The eigenvalue are $\lambda_1 = \sqrt{5}, \lambda_2 = -\sqrt{5}$ and so

$$\begin{bmatrix} 1 & \sqrt{5} \\ 1 & -\sqrt{5} \end{bmatrix} \begin{bmatrix} c_0 \\ c_1 \end{bmatrix} = \begin{bmatrix} e^{\sqrt{5}} \\ e^{-\sqrt{5}} \end{bmatrix} \Rightarrow c_0 = \frac{e^{\sqrt{5}} + e^{-\sqrt{5}}}{2}, \quad c_1 = \frac{e^{\sqrt{5}} - e^{-\sqrt{5}}}{2\sqrt{5}}$$

Therefore

$$e^A = \frac{e^{\sqrt{5}} + e^{-\sqrt{5}}}{2} \begin{bmatrix} 1 & 0 \\ 0 & 1 \end{bmatrix} + \frac{e^{\sqrt{5}} - e^{-\sqrt{5}}}{2\sqrt{5}} \begin{bmatrix} 1 & 2 \\ 2 & -1 \end{bmatrix}$$

$$= \begin{bmatrix} \frac{e^{\sqrt{5}}}{2}\left(1 + \frac{1}{\sqrt{5}}\right) + \frac{e^{-\sqrt{5}}}{2}\left(1 - \frac{1}{\sqrt{5}}\right) & \frac{e^{\sqrt{5}} - e^{-\sqrt{5}}}{\sqrt{5}} \\ \frac{e^{\sqrt{5}} - e^{-\sqrt{5}}}{\sqrt{5}} & \frac{e^{\sqrt{5}}}{2}\left(1 - \frac{1}{\sqrt{5}}\right) + \frac{e^{-\sqrt{5}}}{2}\left(1 + \frac{1}{\sqrt{5}}\right) \end{bmatrix}$$

The above method seems to fail when A has equal eigenvalues, since the

Vandermonde matrix becomes singular, and the coefficients $c_0, \ldots, c_{n-1}$ cannot be easily obtained. However, for the case where A is semi-simple, the minimal polynomial is of degree less than the characteristic polynomial and is equal exactly to $\prod_{i=1}^{s} (\lambda - \lambda_i)$, with s being the number of distinct eigenvalues of A. In that case A^s can be given in terms of $A_1, \ldots, A^{s-1}$ and the Vandermonde matrix will contain only the distinct eigenvalues, and will therefore be smaller in size. The reader can prove for himself that if $\lambda_1, \lambda_2, \ldots, \lambda_s$ are the distinct eigenvalues of A, then the coefficients $c_0, c_1, \ldots, c_{s-1}$ will be given by

$$\begin{bmatrix} 1 & \lambda_1 & \cdots & \lambda_1^{s-1} \\ 1 & \lambda_2 & \cdots & \lambda_2^{s-1} \\ \vdots & & & \\ 1 & \lambda_s & \cdots & \lambda_s^{s-1} \end{bmatrix} \begin{bmatrix} c_0 \\ c_1 \\ \vdots \\ c_{s-1} \end{bmatrix} \begin{bmatrix} e^{\lambda_1} \\ e^{\lambda_2} \\ \vdots \\ e^{\lambda_s} \end{bmatrix}$$

Example. Calculate e^{At}, for

$$A = \begin{bmatrix} 5 & -2 & -4 \\ -2 & 2 & 2 \\ -4 & 2 & 5 \end{bmatrix}$$

The eigenvalues were found to be 10, 1, 1; hence the coefficients $c_0(t)$ and $c_1(t)$ are calculated from solving the equations

$$\begin{bmatrix} 1 & 1 \\ 1 & 10 \end{bmatrix} \begin{bmatrix} c_0(t) \\ c_1(t) \end{bmatrix} = \begin{bmatrix} e^t \\ e^{10t} \end{bmatrix}$$

giving

$$c_0 = \frac{e^{10t} - 10e^t}{-9}, \quad c_1 = \frac{e^t - e^{10t}}{-9}$$

Therefore

$$e^{At} = -\frac{1}{9}(e^{10t} - 10e^t)I + \frac{1}{9}(e^{10t} - e^t) \begin{bmatrix} 5 & -2 & -4 \\ -2 & 2 & 2 \\ -4 & 2 & 5 \end{bmatrix}$$

$$= \frac{1}{9} \begin{bmatrix} 4e^{10t} + 5e^t & 2e^t - 2e^{10t} & 4e^t - 4e^{10t} \\ 2e^t - 2e^{10t} & e^{10t} + 8e^t & 2e^{10t} - 2e^t \\ 4e^t - 4e^{10t} & 2e^{10t} - 2e^t & 4e^{10t} + 5e^t \end{bmatrix}$$

Finally, if A is non-semi-simple, the above technique fails. The reason is that the minimal polynomial is sometimes equal to the characteristic polynomial with the existence of multiplicity in the eigenvalues. The Vandermonde matrix will be singular with no possibility of making it nonsingular. The method utilized will then be as follows: let $u^1, \tilde{u}^2, \ldots, \tilde{u}^m, u^{m+1}, \ldots, u^n$ be the generalized eigenvectors of A with $u^1, \tilde{u}^2, \ldots, \tilde{u}^m$ corresponding to a non-semi-simple eigenvalue λ with multiplicity equal to m. e^A will take the general form:

$$e^A = c_0 I + c_1 A + \cdots + c_{n-1} A^{n-1}$$

Post-multiplying with $u^{m+1}, \ldots, u^n$ gives

$$\begin{aligned} e^{\lambda_{m+1}} &= c_0 + c_1\lambda_{m+1} + \cdots + c_{n-1}\lambda_{m+1}^{n-1} \\ &\vdots \\ e^{\lambda_n} &= c_0 + c_1\lambda_n + \cdots + c_{n-1}\lambda_n^{n-1} \end{aligned}$$

Whereas by post-multiplying with $\tilde{u}^2$, while using the relations

$$\begin{aligned} Au^1 &= \lambda u^1 \\ A\tilde{u}^2 &= \lambda \tilde{u}^2 + u^1 \\ &\vdots \\ A\tilde{u}^m &= \lambda \tilde{u}^m + \tilde{u}^{m-1}, \end{aligned}$$

and assuming that A is non-derogatory without loss of generality, we obtain

$$e^A \tilde{u}^2 = e^\lambda(\tilde{u}^2 + u^1) = c_0\tilde{u}^2 + c_1(\lambda\tilde{u}^2 + u^1) + c_2(\lambda^2\tilde{u}^2 + 2\lambda u^1) + \cdots + \\ + c_{n-1}(\lambda^{n-1}\tilde{u}^2 + (n-1)\lambda^{n-2}u^1)$$

But as u^1 and $\tilde{u}^2$ are linearly independent, their coefficients in the above equation vanish and we obtain

$$\begin{aligned} e^\lambda &= c_0 + c_1\lambda + \cdots + c_{n-1}\lambda^{n-1} \\ e^\lambda &= c_1 + 2c_2\lambda + \cdots + (n-1)c_{n-1}\lambda^{n-2} \end{aligned}$$

And if the same procedure is repeated for $\tilde{u}^3, \ldots, \tilde{u}^m$ we obtain similarly

$$\begin{aligned} e^\lambda &= c_0 + c_1\lambda + \cdots + c_{n-1}\lambda^{n-1} \\ e^\lambda &= c_1 + 2c_2\lambda + \cdots + (n-1)c_{n-1}\lambda^{n-2} \\ e^\lambda &= 2c_2 + 6c_3\lambda + \cdots + (n-1)(n-2)c_{n-1}\lambda^{n-3} \\ &\vdots \\ e^\lambda &= (m-1)!c_{m-1} + \cdots + P_{m-1}^{n-2} c_{n-2}\lambda^{n-m-1} + P_{m-1}^{n-1} c_{n-1}\lambda^{n-m} \end{aligned}$$

The above m equations with the remaining $n - m$ equations concerning the distinct eigenvalues can be organized in the following form:

$$\begin{bmatrix} 1 & \lambda_{m+1} & \lambda_{m+1}^2 & \cdots & & & & \lambda_{m+1}^{n-1} \\ 1 & \lambda_{m+2} & \lambda_{m+2}^2 & \cdots & & & & \vdots \\ \vdots & \vdots & \vdots & & & & & \\ 1 & \lambda_n & \lambda_n^2 & & & & & \lambda_n^{n-1} \\ 1 & \lambda & \lambda^2 & & & & & \lambda^{n-1} \\ 0 & 1 & 2\lambda & & & & & (n-1)\lambda^{n-2} \\ 0 & 0 & 2 & & & & & (n-1)(n-2)\lambda^{n-3} \\ \vdots & \vdots & \vdots & & & & & \vdots \\ 0 & 0 & 0 & \cdots & (m-1)! & \cdots & P_{m-1}^{n-2}\lambda^{n-m-1} & P_{m-1}^{n-1}\lambda^{n-m} \end{bmatrix} \begin{bmatrix} c_0 \\ c_1 \\ \vdots \\ \\ \\ \\ c_{n-1} \end{bmatrix} = \begin{bmatrix} e^{\lambda_{m+1}} \\ e^{\lambda_{m+2}} \\ \vdots \\ e^{\lambda_n} \\ e^{\lambda} \\ \vdots \\ e^{\lambda} \end{bmatrix}$$

The reader can verify that the matrix in the left-hand side is nonsingular. Therefore the coefficients $c_0, c_1, \ldots, c_{n-1}$ can be determined uniquely. Note that if the eigenvalues $\lambda_{m+1}, \ldots, \lambda_n$ are not distinct, but only semi-simple, we shall disregard their multiplicities and keep only the distinct ones; of course the minimal polynomial of A will be of order less than n by the number of superfluous equal eigenvalues of $\lambda_{m+1}, \ldots, \lambda_n$.

Example. Calculate e^{At}, if $A = \begin{bmatrix} 2 & -1 \\ 1 & 0 \end{bmatrix}$

The eigenvalues are $\lambda_1 = \lambda_2 = 1$, and the matrix is non-semi-simple; hence $c_0(t), c_1(t)$ are calculated from

$$\begin{bmatrix} 1 & 1 \\ 0 & 1 \end{bmatrix} \begin{bmatrix} c_0(t) \\ c_1(t) \end{bmatrix} = \begin{bmatrix} e^t \\ te^t \end{bmatrix}$$

giving

$$c_1 = te^t, \quad c_0 = e^t - te^t$$

$$e^{At} = (e^t - te^t)\begin{bmatrix} 1 & 0 \\ 0 & 1 \end{bmatrix} + te^t \begin{bmatrix} 2 & -1 \\ 1 & 0 \end{bmatrix} = \begin{bmatrix} e^t + te^t & -te^t \\ te^t & e^t - te^t \end{bmatrix}$$

Next we explain how the spectral representation of a matrix helps in computing matrix functions. Recalling the relation of Section 4.10, i.e.

$$A = \sum_{i=1}^{n} \lambda_i E_i + \sum_{i=1}^{m-1} u^i \rangle\langle v^{i+1}$$

of a general non-semi-simple matrix A, we can deduce some popular matrix functions. To compute A^{-1} we pre-multiply the relation

$$I = \sum_{i=1}^{n} u^i \rangle\langle v^i$$

by A^{-1}. For A semi-simple the reader can check that

$$A^{-1} = \sum_{i=1}^{n} \frac{E_i}{\lambda_i}$$

But for A non-semi-simple, the above formula needs further adjustment. Let us go back to the eigenvalue problem of the generalized eigenvectors of A. We called them in Section 4.10 $u^1, u^2, \ldots, u^m, u^{m+1}, \ldots, u^n$, where the first m vectors are the generalized eigenvectors corresponding to the non-semi-simple eigenvalue λ. Therefore we have:

$$\begin{aligned} Au^1 &= \lambda u^1 \\ Au^2 &= \lambda u^2 + u^1 \\ &\vdots \\ Au^m &= \lambda u^m + u^{m-1}, \end{aligned}$$

from which we obtain

$$A^{-1}u^i = \frac{1}{\lambda_i} u^i, \quad i = 1, m+1, \ldots, n$$

$$A^{-1}u^2 = \frac{u^2}{\lambda} - \frac{A^{-1}u^1}{\lambda} = \frac{u^2}{\lambda} - \frac{u^1}{\lambda^2}$$

$$A^{-1}u^3 = \frac{u^3}{\lambda} - \frac{A^{-1}u^2}{\lambda} = \frac{u^3}{\lambda} - \frac{u^2}{\lambda^2} + \frac{u^1}{\lambda^3}$$

$$\vdots$$

$$A^{-1}u^m = \frac{u^m}{\lambda} - \frac{u^{m-1}}{\lambda^2} + \frac{u^{m-2}}{\lambda^3} - \cdots + (-)^{m-1} \frac{u^1}{\lambda^m}$$

Now pre-multiply by A^{-1} the relation

$$I = \sum_{i=1}^{n} u^i \rangle\langle v^i$$

As before, we obtain

$$A^{-1} = \sum_{i=1}^{n} \frac{E_i}{\lambda_i} - \sum_{i=1}^{m-1} \frac{u^i\rangle\langle v^{i+1}}{\lambda^2} + \sum_{i=1}^{m-2} \frac{u^i\rangle\langle v^{i+2}}{\lambda^3} + \cdots$$

$$+ (-)^{m-1} \frac{u^1\rangle\langle v^m}{\lambda^m}$$

The reader may find it absurd to compute A^{-1} this way, but he can check that A^{-1} represented in this form is no more than

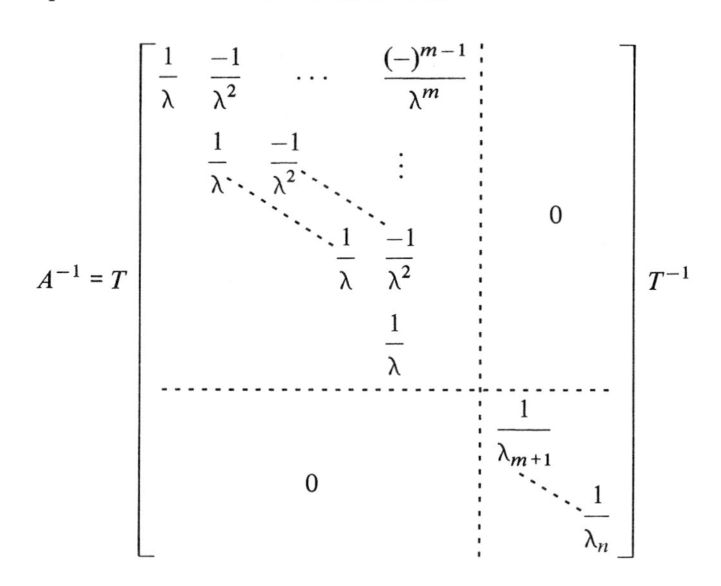

We will not exhaust the reader by computing e^{At} as before and we give the formula direct:

$$e^{At} = T \left[\begin{array}{ccccc:ccc} e^{\lambda t} & te^{\lambda t} & \frac{t^2 e^{\lambda t}}{2!} & \cdots & \frac{t^{m-1} e^{\lambda t}}{(m-1)!} & & & \\ & e^{\lambda t} & \ddots & & \vdots & & 0 & \\ & & \ddots & e^{\lambda t} & te^{\lambda t} & & & \\ & & & & e^{\lambda t} & & & \\ \hdashline & & & & & e^{\lambda_{m+1} t} & & \\ & & 0 & & & & \ddots & \\ & & & & & & & e^{\lambda_n t} \end{array}\right] T^{-1}$$

One other important matrix function is A^k. The reader is asked to calculate A^k when k is a positive integer. However we give here a formula for $\sqrt{A}$, which is an interesting matrix function:

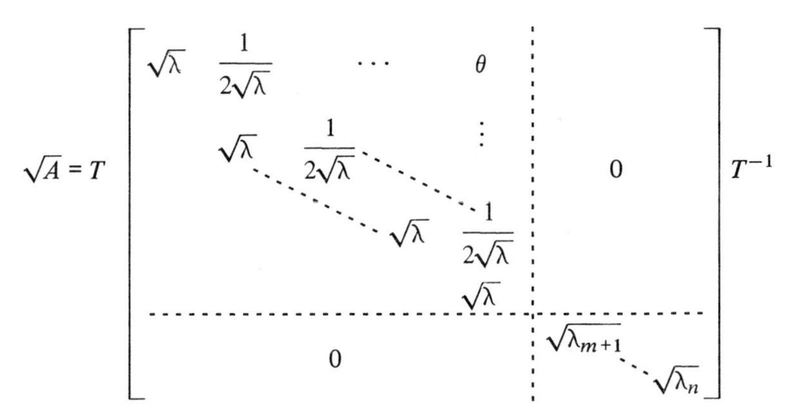

with

$$\theta = \frac{(-)^m(2m-5)(2m-7)\ldots 5 \times 3 \times 1}{(m-1)!2^{m-1}} \lambda^{-(2m-3)/2}$$

The reader is asked to compute $\sin A$, $\cos A$, $\ln(I+A)$, $\sinh A$, $\cosh A$, $(\lambda I - A)^{-1}$.

Exercises 5.6

1. Show that for any square matrix A:

(a) $e^{iA} = \cos A + i \sin A, \quad i = \sqrt{-1}$
(b) $\sin 2A = 2 \sin A \cdot \cos A$
(c) $\cos^2 A + \sin^2 A = I$

2. Show that $| e^A | = e^{\mathrm{tr}\, A}$.
3. Show that $\| e^A \| \leqslant e^{\| A \|}$.
4. A matrix A is called nilpotent if $A^k = 0$, for some positive integer k; show that $\sum_{i=1}^{m} A^i = I(I-A)^{-1}, m \geqslant k-1.$
5. A matrix A is called idempotent if $A^k = A$, for some positive integer k; find $\sum_{i=1}^{k} A^i.$
6. Evaluate $\sin At$, if $A = \begin{bmatrix} 5 & -2 & -4 \\ -2 & 2 & 2 \\ -4 & 2 & 5 \end{bmatrix}$
7. Evaluate $\sqrt{A}$, if A is as above.
8. Show that A^m, with m very large, is given by $A^m \approx \hat{\lambda}^m E$, with $\hat{\lambda}$ the maximum eigenvalue of A in magnitude and E the eigenprojection corresponding to $\hat{\lambda}$. Should this result hold true only if $\hat{\lambda}$ is a semi-simple eigenvalue of A? Hence obtain A^{20} for the matrix given above.

9. Show that the solution of the difference equation

$$x(k+1) = Ax(k) + y(k), \quad \text{where } k \text{ is an integer}$$

is given by

$$x(k) = A^k x(0) + \sum_{j=1}^{n} A^{n-j} y(j-1)$$

If y = a constant vector E, and the eigenvalues of A are less than unity, show that $\lim_{k\to\infty} x(k) = (I - A)^{-1}E$.

10. If A is semi-simple, show that in general $F(A) = \sum_i f(\lambda_i)E_i$.

11. If $AB = BA$, show that $\sin A \sin B = \sin B \sin A$. In general, show that $F_1(A)F_2(B) = F_2(B)F_1(A)$. *Hint*: A and B have the same eigenprojection; use the spectral representation $F(A) = \sum_i f(\lambda_i)E_i$.

12. Show that $e^A \cdot e^B = e^{A+B}$ if $AB = BA$.
13. Prove the following theorem: two functions of the same matrix commute.
14. Show that $de^{At}/dt = Ae^{At}$ and that

$$\int_0^t e^{At}\, dt = A^{-1}e^{At} - A^{-1}.$$

What happens if A is singular?

15. Find $\sqrt{A}$ if $A = \begin{bmatrix} 1 & 1 \\ 0 & 1 \end{bmatrix}$.

16. Show that the function $\sqrt{A}$ is not unique.
17. Show that, if the eigenvalues of A have negative real parts, $\lim_{t\to\infty} e^{At} = 0$.

18. If $AB = BA$, show that the eigenvalues of $f(A, B)$ are $f(\lambda_i, \mu_i)$ where λ_i and μ_i are respectively the eigenvalues of A and B arranged in a fixed order, independent of f.
19. If the Lyapunov equation $AX + XB = C$, where A, B, C and X are square matrices, is written in the form $(A \otimes I + I \otimes B^T)x = c$, where x and c are vectors containing all the elements of X and C in a suitable arrangement, this defines the *Kronecker product* $\otimes$ *of two matrices*. In general if R and P are two square matrices of order m and n, then $R \otimes P$ is of order (m, n). Show that

(a) $I \otimes B = \text{quasidiag}(B, B, \ldots, B)$
(b) $(A \otimes B) \otimes C = A \otimes (B \otimes C)$
(c) $(A + B) \otimes (C + D) = A \otimes C + A \otimes D + B \otimes C + B \otimes D$
(d) $(A \otimes B)(C \otimes D) = (AC) \otimes (BD)$
(e) $\lambda(A \otimes B) = \lambda_i(A)\mu_j(B)$. *Hint*: write $Ax = \lambda x$. $By = \mu y$, then arrange a suitable multiplication; from which you can also obtain the eigenvector of $A \otimes B$
(f) $\text{tr}(A \otimes B) = (\text{tr } A)(\text{tr } B)$
(g) $|A \otimes B| = |B \otimes A| = |A|^n |B|^m$, if A and B are of order m and n respectively
(h) $(A \otimes B)^{-1} = A^{-1} \otimes B^{-1}$
(i) $e^{A \otimes I + I \otimes B} = e^A \otimes e^B$

(j) $\lambda(A \otimes I + I \otimes B) = \lambda_i(A) + \mu_j(B)$. *Hint*: write $(I + \epsilon A) \otimes (I + \epsilon B) = I \otimes I + \epsilon(A \otimes I + I \otimes B) + \epsilon^2 A \otimes B$, whose eigenvalues are $(1 + \epsilon\lambda_i)(1 + \epsilon\mu_j)$

The last result is useful for establishing the existence of a solution of the Lyapunov equation. By writing the equation $AX + XB = C$ in the form $(A \otimes I + I \otimes B^T)x = c$, we realize that it has a solution X for all C if and only if $\lambda_i(A) + \mu_j(B) \neq 0$. The Lyapunov equation appears frequently in the theory of linear dynamical systems associated with problems of control and stability. The equations can be solved using available algorithms for linear equations; except that when A and B are large, these algorithms tend to be time and storage consuming. For if A and B are of order n, the number of equations will rise to n^2. Bartels and Stewart (1972) drew an equivalence between the above problem and solving the equations $(T^{-1}AT)(T^{-1}XS) + (T^{-1}XS)(S^{-1}BS) = T^{-1}CS$. As an exercise, write the steps in solving these equations. Further modification to the problem was made by Golub *et al.* (1979), in which they also studied $AXM + X = C$ and $AXM + LXB = C$. As for the Riccati equation $F^TX + XF - XGX + H = 0$, which appears in many areas of applied mathematics, the reader is referred to Laub (1979), who surveyed available methods of solution. For further reading about the Kronecker product and related fields, refer to Bellman (1970), Chapter 12, and Pease (1965), Chapter 14.

20. Prove the following lemma:

$$\| AX + XB \| \geqslant \| X \| \frac{\min_{i,j} | \lambda_i + \mu_j |}{\mathscr{K}(A)\,\mathscr{K}(B)},$$

where λ and μ are the eigenvalues of A and B and $\mathscr{K}(A)$ and $\mathscr{K}(B)$ are respectively their spectral condition numbers. *Hint*: write $T^{-1}(AX + XB)S = T^{-1}ATT^{-1}XS + T^{-1}XSS^{-1}BS$.

5.7. Application to ordinary differential equations

Ordinary differential equations appear in the formulation of many physical problems. For example the differential equations of any passive linear electrical circuit composed of inductors, capacitors and resistors can be formulated in the following form:

$$\frac{\mathrm{d}x}{\mathrm{d}t} = A(t)x + Bu(t)$$

where x is the state vector containing the state variables of the circuit (see Macfarlane (1970), Chapter 5). They can be the independent inductors' currents and independent capacitors' voltages and the number of these equals the total number of independent initial conditions. A is the *system state matrix*, whose elements are functions of the circuit elements: B is the *input matrix*; again it is dependent on the circuit elements: u is the *control vector*, which contains the current and voltage sources. If A is constant, i.e. the circuit elements are time-invariant, we call the differential equation linear with constant coefficients; whereas if the elements are time-dependent, A will be a function of time and the differential equation will then be a time-varying one.

Example. Write the differential equation of the following circuit:

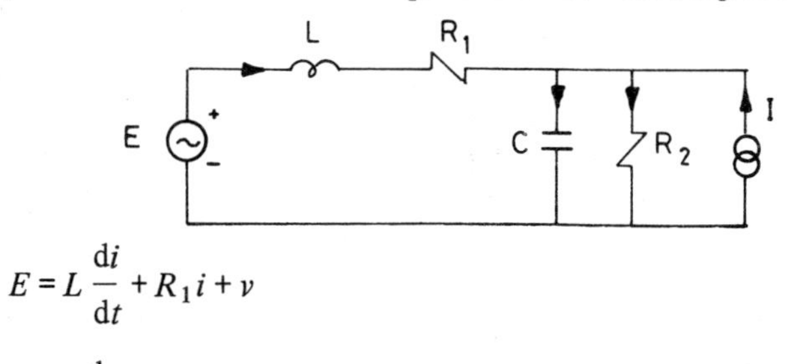

$$E = L\frac{\mathrm{d}i}{\mathrm{d}t} + R_1 i + v$$

$$i = C\frac{\mathrm{d}v}{\mathrm{d}t} + \frac{v}{R_2} - I$$

where i and v are respectively the current passing through the inductor and the voltage drop across the capacitor. The above equations can be formulated in the matrix form

$$\frac{\mathrm{d}}{\mathrm{d}t}\begin{bmatrix} i \\ v \end{bmatrix} = \begin{bmatrix} \frac{-R_1}{L} & -\frac{1}{L} \\ \frac{1}{C} & \frac{-1}{R_2 C} \end{bmatrix}\begin{bmatrix} i \\ v \end{bmatrix} + \begin{bmatrix} \frac{E}{L} \\ \frac{I}{C} \end{bmatrix}$$

Another example is the differential equation of a mechanical system composed of a set of masses, dash-pots and springs connected together; it takes the form

$$M\frac{\mathrm{d}^2 q}{\mathrm{d}t^2} + B\frac{\mathrm{d}q}{\mathrm{d}t} + Kq = F(t)$$

where M, B and K are respectively the mass matrix, damping matrix and stiffness matrix, and are in general positive semidefinite; $F(t)$ is the forcing vector; and $q(t)$ is a vector representing the state variables, which may be displacements of the system. The dimension of $q(t)$ is equal to the number of degrees of freedom of the system.

Example. Write the differential equation of the mechanical system with two degrees of freedom:

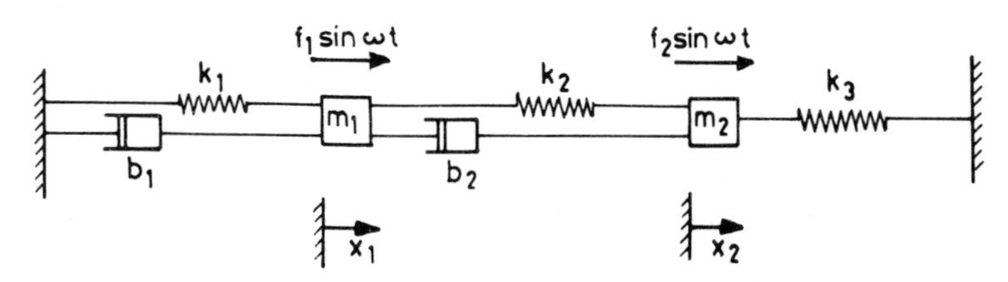

$$F_1 \sin \omega t - k_1 x_1 - b_1 \frac{dx_1}{dt} = m_1 \frac{d^2 x_1}{dt^2} + k_2(x_1 - x_2) + b_2 \frac{d}{dt}(x_1 - x_2)$$

$$F_2 \sin \omega t - k_3 x_2 = m_2 \frac{d^2 x_2}{dt^2} + k_2(x_2 - x_1) + b_2 \frac{d}{dt}(x_2 - x_1)$$

The above equations can be organized in the following form:

$$\begin{bmatrix} m_1 & 0 \\ 0 & m_2 \end{bmatrix} \begin{bmatrix} \dfrac{d^2 x_1}{dt^2} \\ \dfrac{d^2 x_2}{dt^2} \end{bmatrix} + \begin{bmatrix} b_1 + b_2 & -b_2 \\ -b_2 & b_2 \end{bmatrix} \begin{bmatrix} \dfrac{dx_1}{dt} \\ \dfrac{dx_2}{dt} \end{bmatrix} +$$

$$+ \begin{bmatrix} k_1 + k_2 & -k_2 \\ -k_2 & k_2 + k_3 \end{bmatrix} \begin{bmatrix} x_1 \\ x_2 \end{bmatrix} = \begin{bmatrix} F_1 \sin \omega t \\ F_2 \sin \omega t \end{bmatrix}$$

The above second-order differential equations can be transformed into first-order ones, by making the proper substitutions

$$\frac{dx_1}{dt} = \theta_1, \quad \frac{dx_2}{dt} = \theta_2$$

The equations will then be of first-order, but the state matrix will be of fourth-order.

In general any linear differential equation of nth order can be transformed into one of first-order. For, let the differential equation

$$A_n \frac{d^n x}{dt^n} + A_{n-1} \frac{d^{n-1} x}{dt^{n-1}} + \cdots + A_1 \frac{dx}{dt} + A_0 x = F(t)$$

where A_n is assumed nonsingular for the moment. Then by making the following substitutions

$$\frac{dx}{dt} = x^1$$

$$\frac{dx^1}{dt} = x^2$$

$$\vdots$$

$$\frac{dx^{n-2}}{dt} = x^{n-1}$$

it will read

$$\frac{d}{dt}\begin{bmatrix} x \\ x^1 \\ \vdots \\ x^{n-1} \end{bmatrix} = \begin{bmatrix} 0 & I & 0 & \cdots & 0 \\ 0 & 0 & I & \ddots & 0 \\ \vdots & & & \ddots & \vdots \\ 0 & & & 0 & I \\ -A_n^{-1}A_0 & & \cdots & -A_n^{-1}A_{n-2} & -A_n^{-1}A_{n-1} \end{bmatrix} \begin{bmatrix} x \\ x^1 \\ \vdots \\ x^{n-1} \end{bmatrix} + \begin{bmatrix} 0 \\ 0 \\ \vdots \\ A_n^{-1}F(t) \end{bmatrix}$$

Note that the matrix on the right-hand side is the companion matrix of the differential equation. If the matrices $A_0, A_1, \ldots, A_n$ are of order (m, m), the companion matrix will be of order (m, n).

Example. Transform the differential equation

$$\begin{bmatrix} 1 & 0 \\ -1 & 2 \end{bmatrix} \begin{bmatrix} \dfrac{d^2x_1}{dt^2} \\ \dfrac{d^2x_2}{dt^2} \end{bmatrix} + \begin{bmatrix} 1 & -1 \\ 2 & 3 \end{bmatrix} \begin{bmatrix} \dfrac{dx_1}{dt} \\ \dfrac{dx_2}{dt} \end{bmatrix} + \begin{bmatrix} -3 & 0 \\ 1 & 2 \end{bmatrix} \begin{bmatrix} x_1 \\ x_2 \end{bmatrix} = \begin{bmatrix} \sin t \\ e^t \end{bmatrix}$$

into a first-order one.

Substituting

$$\frac{dx_1}{dt} = y_1, \quad \frac{dx_2}{dt} = y_2$$

we obtain

$$\begin{bmatrix} 1 & 0 & 0 & 0 \\ -1 & 2 & 0 & 0 \\ 0 & 0 & 1 & 0 \\ 0 & 0 & 0 & 1 \end{bmatrix} \begin{bmatrix} \dfrac{dy_1}{dt} \\ \dfrac{dy_2}{dt} \\ \dfrac{dx_1}{dt} \\ \dfrac{dx_2}{dt} \end{bmatrix} = \begin{bmatrix} -1 & 1 & 3 & 0 \\ -2 & -3 & -1 & -2 \\ 1 & 0 & 0 & 0 \\ 0 & 1 & 0 & 0 \end{bmatrix} \begin{bmatrix} y_1 \\ y_2 \\ x_1 \\ x_2 \end{bmatrix} + \begin{bmatrix} \sin t \\ e^t \\ 0 \\ 0 \end{bmatrix}$$

i.e.

$$\frac{\mathrm{d}}{\mathrm{d}t}\begin{bmatrix} y_1 \\ y_2 \\ x_1 \\ x_2 \end{bmatrix} = \begin{bmatrix} -1 & 1 & 3 & 0 \\ -3/2 & -1 & 1 & -1 \\ 1 & 0 & 0 & 0 \\ 0 & 1 & 0 & 0 \end{bmatrix}\begin{bmatrix} y_1 \\ y_2 \\ x_1 \\ x_2 \end{bmatrix} + \begin{bmatrix} \sin t \\ \tfrac{1}{2}(\sin t + e^t) \\ 0 \\ 0 \end{bmatrix}$$

Example. Transform the differential equation

$$\frac{\mathrm{d}^3y}{\mathrm{d}x^3} + \sin x\,\frac{\mathrm{d}^2y}{\mathrm{d}x^2} + x\,\frac{\mathrm{d}y}{\mathrm{d}x} + 2y = \cos x$$

into a first-order one.

Substituting

$$\frac{\mathrm{d}y}{\mathrm{d}x} = y_1, \quad \frac{\mathrm{d}y_1}{\mathrm{d}x} = y_2$$

we obtain the first-order equation

$$\frac{\mathrm{d}}{\mathrm{d}x}\begin{bmatrix} y_2 \\ y_1 \\ y \end{bmatrix} = \begin{bmatrix} -\sin x & -x & -2 \\ 1 & 0 & 0 \\ 0 & 1 & 0 \end{bmatrix}\begin{bmatrix} y_2 \\ y_1 \\ y \end{bmatrix} + \begin{bmatrix} \cos x \\ 0 \\ 0 \end{bmatrix}$$

This is a time-varying differential equation whose initial conditions are $y(0), y_1(0)$ and $y_2(0)$, which are $y(0), \left.\frac{\mathrm{d}y}{\mathrm{d}x}\right|_{x=0}$ and $\left.\frac{\mathrm{d}^2y}{\mathrm{d}x^2}\right|_{x=0}$.

It seems that solving any linear differential equation of any order boils down to the solution of a first-order linear differential equation. So we start by solving a linear time-invariant one:

$$\frac{\mathrm{d}x}{\mathrm{d}t} = Ax + u(t)$$

and consider the most general case when A is non-semi-simple. Substituting

$$x = Tx', \quad u = Tu'$$

we obtain

$$T\frac{\mathrm{d}x'}{\mathrm{d}t} = ATx' + Tu'$$

i.e.

$$\frac{\mathrm{d}x'}{\mathrm{d}t} = T^{-1}ATx' + u'$$

and if T is chosen to be the modal matrix of A, we obtain

$$\frac{d}{dt}\begin{bmatrix} x'_1 \\ x'_2 \\ \vdots \\ x'_n \end{bmatrix} = \left[\begin{array}{cccc|ccc} \lambda & \delta & & & & & \\ & \lambda & \delta & & & 0 & \\ & & \ddots & \delta & & & \\ & & & \lambda & & & \\ \hline & & & & \lambda_{m+1} & & \\ & 0 & & & & \ddots & \\ & & & & & & \lambda_n \end{array}\right]\begin{bmatrix} x'_1 \\ x'_2 \\ \vdots \\ x'_n \end{bmatrix} + \begin{bmatrix} u'_1 \\ u'_2 \\ \vdots \\ u'_n \end{bmatrix}$$

where the matrix on the right-hand side is the Jordan form for A. The solution of the above equation is obtained by solving each differential equation for $x'_i, i = 1, \ldots, n$. For a semi-simple eigenvalue the scalar differential equation is

$$\frac{dx'_i}{dt} = \lambda_i x'_i + u'_i(t), \quad i = m + 1, \ldots, n$$

whose solution can be written directly:

$$x'_i(t) = e^{\lambda_i t} x'_i(0) + e^{\lambda_i t} \int_0^t e^{-\lambda_i \tau} u'_i(\tau)\, d\tau$$

And for a non-semi-simple eigenvalue, the scalar differential equations read

$$\frac{dx'_m}{dt} = \lambda x'_m + u'_m(t)$$

$$\frac{dx'_{m-1}}{dt} = \lambda x'_{m-1} + x'_m + u'_{m-1}(t)$$

$$\vdots$$

$$\frac{dx'_1}{dt} = \lambda x'_1 + x'_2 + u'_1(t)$$

Starting from the first equation and moving downwards, we obtain

$$x'_m(t) = e^{\lambda t} x'_m(0) + e^{\lambda t} \int_0^t e^{-\lambda \tau} u'_m(\tau)\, d\tau$$

Substituting for $x'_m(t)$ in the second equation and solving, we obtain

$$x'_{m-1}(t) = e^{\lambda t} x'_{m-1}(0) + e^{\lambda t} \int_0^t e^{-\lambda \tau}\Big(e^{\lambda \tau} x'_m(0) +$$

$$+ e^{\lambda \tau} \int_0^\tau e^{-\lambda z} u'_m(z)\, dz + u'_{m-1}(\tau)\Big)\, d\tau$$

$$= e^{\lambda t}x'_{m-1}(0) + e^{\lambda t}\int_0^t e^{-\lambda\tau}u'_{m-1}(\tau)\,\mathrm{d}\tau + te^{\lambda t}x'_m(0)$$

$$-e^{\lambda t}\int_0^t \tau e^{-\lambda\tau}u'_m(\tau)\,\mathrm{d}\tau + te^{\lambda t}\int_0^t e^{-\lambda\tau}u'_m(\tau)\,\mathrm{d}\tau$$

The process is repeated until we reach $x'_1(t)$. The solution $x'(t)$ can be written in the following manner:

$$\begin{bmatrix} x'_1(t) \\ x'_2(t) \\ \vdots \\ x'_n(t) \end{bmatrix} = \left[\begin{array}{cccc:ccc} e^{\lambda t} & te^{\lambda t} & \cdots & \dfrac{t^{m-1}e^{\lambda t}}{(m-1)!} & & & \\ & e^{\lambda t} & \ddots & \vdots & & 0 & \\ & & \ddots & te^{\lambda t} & & & \\ & & & e^{\lambda t} & & & \\ \hdashline & & & & e^{\lambda_{m+1}t} & & \\ & 0 & & & & \ddots & \\ & & & & & & e^{\lambda_n t} \end{array}\right] \begin{bmatrix} x'_1(0) \\ x'_2(0) \\ \vdots \\ x'_n(0) \end{bmatrix} +$$

$$+ \left[\begin{array}{cccc:ccc} e^{\lambda t} & te^{\lambda t} & \cdots & \dfrac{t^{m-1}e^{\lambda t}}{(m-1)!} & & & \\ & e^{\lambda t} & \ddots & \vdots & & 0 & \\ & & & e^{\lambda t} & & & \\ \hdashline & & & & e^{\lambda_{m+1}t} & & \\ & 0 & & & & \ddots & e^{\lambda_n t} \end{array}\right] \times$$

$$\times \int_0^t \left[\begin{array}{cccc:ccc} e^{-\lambda\tau} & -\tau e^{-\lambda\tau} & \cdots & (-)^{m-1}\dfrac{\tau^{m-1}e^{-\lambda\tau}}{(m-1)!} & & & \\ & e^{-\lambda\tau} & \ddots & \vdots & & 0 & \\ & & & e^{-\lambda\tau} & & & \\ \hdashline & & & & e^{-\lambda_{m+1}\tau} & & \\ & 0 & & & & \ddots & e^{\lambda_n\tau} \end{array}\right] \begin{bmatrix} u'_1(\tau) \\ u'_2(\tau) \\ \vdots \\ u'_n(\tau) \end{bmatrix} \mathrm{d}\tau$$

Substituting $x' = T^{-1}x$, $u' = T^{-1}u$, we obtain

$$x(t) = e^{At}x(0) + e^{At}\int_0^t e^{-A\tau}u(\tau)\,\mathrm{d}\tau$$

which is the solution of the first-order linear differential equation with constant coefficients, and e^{At} can be calculated as before.

Example. Solve the differential equations

$$\frac{\mathrm{d}}{\mathrm{d}t}\begin{bmatrix} y \\ x \end{bmatrix} = \begin{bmatrix} 2 & -1 \\ 1 & 0 \end{bmatrix}\begin{bmatrix} y \\ x \end{bmatrix} + \begin{bmatrix} t \\ 0 \end{bmatrix}, \quad x(0) = 1, \quad y(0) = -1$$

e^{At} is calculated as before:

$$e^{At} = \begin{bmatrix} te^t + e^t & -te^t \\ te^t & e^t - te^t \end{bmatrix}$$

and we obtain

$$\begin{bmatrix} y(t) \\ x(t) \end{bmatrix} = \begin{bmatrix} te^t + e^t & -te^t \\ te^t & e^t - te^t \end{bmatrix}\begin{bmatrix} -1 \\ 1 \end{bmatrix} +$$

$$+\begin{bmatrix} e^t + te^t & -te^t \\ te^t & e^t - te^t \end{bmatrix}\int_0^t \begin{bmatrix} -\tau e^{-\tau} + e^{-\tau} & \tau e^{-\tau} \\ -\tau e^{-\tau} & e^{-\tau} + \tau e^{-\tau} \end{bmatrix}\begin{bmatrix} \tau \\ 0 \end{bmatrix} \mathrm{d}\tau$$

Integrating, we obtain

$$x(t) = -te^t - e^t + 2 + t$$
$$y(t) = -te^t - 2e^t + 1$$

Example. Solve the differential equations

$$\frac{\mathrm{d}x}{\mathrm{d}t} = 5x - 2y - 4z$$

$$\frac{\mathrm{d}y}{\mathrm{d}t} = -2x + 2y + 2z \qquad \text{subject to } x(0) = 9, y(0) = -9, z(0) = 18$$

$$\frac{\mathrm{d}z}{\mathrm{d}t} = -4x + 2y + 5z$$

The solution is:

$$\begin{bmatrix} x(t) \\ y(t) \\ z(t) \end{bmatrix} = e^{At}\begin{bmatrix} 9 \\ -9 \\ 18 \end{bmatrix} = \begin{bmatrix} 4e^{10t} + 5e^t & 2e^t - 2e^{10t} & 4e^t - 4e^{10t} \\ 2e^t - 2e^{10t} & e^{10t} + 8e^t & 2e^{10t} - 2e^t \\ 4e^t - 4e^{10t} & 2e^{10t} - 2e^t & 4e^{10t} + 5e^t \end{bmatrix}\begin{bmatrix} 1 \\ -1 \\ 2 \end{bmatrix}$$

giving

$$x(t) = -2e^{10t} + 11e^t$$
$$y(t) = -10e^t + e^{10t}$$
$$z(t) = 16e^t + 2e^{10t}$$

Now we treat the case when A is a function of time, i.e. we consider the differential equation

$$\frac{dx}{dt} = A(t)x + u(t)$$

It may seem that the method of solution of this equation is similar to that when A is a constant matrix, but this is not so, since the eigenvalues of A are functions of time and it is generally difficult to find them. Special cases exist when the eigenvalues can be determined easily as for the case:

$$A = \begin{bmatrix} \sin t & \cos t \\ \cos t & -\sin t \end{bmatrix}$$

However in this case the modal matrix is still time-dependent, and so even if A can be diagonalized, the modified differential equation in $x' = T^{-1}x$ is again time-dependent and perhaps more difficult to solve; this makes us look for another method of solving the equation.

The method which is generally used is to use an iterative method like the following:

$$\int_0^t \frac{dx}{dt}\, dt = \int_0^t [A(\tau)x + u(\tau)]\, d\tau,$$

giving

$$x(t) = x(0) + \int_0^t [A(\tau)x + u(\tau)]\, d\tau$$

$$= x(0) + \int_0^t \left[A(\tau) \left[x(0) + \int_0^\tau [A(s)x + u(s)]\, ds \right] + u(\tau) \right] d\tau$$

And if the iteration is continued we obtain

$$x(t) = \left(I + \int_0^t A(\tau)\, d\tau + \int_0^t A(\tau) \int_0^\tau A(s)\, ds\, d\tau + \right.$$

$$\left. + \int_0^t A(\tau) \int_0^\tau A(s) \int_0^s A(z)\, dz\, ds\, d\tau + \cdots \right) x(0) +$$

$$+ \int_0^t u(\tau)\, d\tau + \int_0^t A(\tau) \int_0^\tau u(s)\, ds\, d\tau +$$

$$+ \int_0^t A(\tau) \int_0^\tau A(s) \int_0^s u(z)\, dz\, ds\, d\tau + \cdots$$

And if we make the definition

$$M(t) = I + \int_0^t A(\tau)\, d\tau + \int_0^t A(\tau) \int_0^\tau A(s)\, ds\, d\tau +$$

$$+ \int_0^t A(\tau) \int_0^\tau A(s) \int_0^s A(z)\, dz\, ds\, d\tau + \cdots$$

then the reader can check that the solution becomes

$$x(t) = M(t)x(0) + M(t) \int_0^t M^{-1}(\tau)u(\tau)\, d\tau$$

Now what is left is to prove that the series $M(t)$ converges and that $M^{-1}(t)$ exists. $M(t)$ is called the *matrizant* of the differential equation. To prove that $M(t)$ converges we take norms

$$\| M(t) \| \leqslant 1 + \int_0^t \| A(\tau) \|\, d\tau + \int_0^t \| A(\tau) \| \int_0^\tau \| A(s) \|\, ds\, d\tau + \cdots$$

and defining

$$\max \| A(\tau) \| = l. \quad \forall \tau \in [0, t]$$

we obtain

$$\| M(t) \| \leqslant 1 + lt + \frac{l^2 t^2}{2!} + \cdots = e^{lt}$$

from which we conclude that the series $M(t)$ is absolutely and uniformly convergent for all values of t.

To show that $M^{-1}(t)$ exists is to show that

$$| M(\tau) | \neq 0, \quad \forall \tau \in [0, t]$$

Now from the definition of $M(t)$ we have $dM(t)/dt = A(t)M(t), M(0) = I$. The reader can prove as an interesting exercise that

$$\frac{d}{dt} | M(t) | = | M(t) | \cdot \operatorname{tr} A(t), \quad | M(0) | = 1$$

i.e.

$$| M(t) | = | M(0) | \, e^{\int_0^t (\operatorname{tr} A(\tau))\, d\tau}$$
$$= e^{\int_0^t (\operatorname{tr} A(\tau))\, d\tau}$$

From which we conclude that

$$| M(\tau) | \neq 0, \quad \forall \tau \in [0, t]$$

Example. Solve the differential equation

$$\frac{d^2y}{dt^2} + ty = 0, \quad y(0) = c_1, \quad \dot{y}(0) = c_2$$

Substituting

$$\frac{dy}{dt} = x,$$

we obtain

$$\frac{d}{dt}\begin{bmatrix} y \\ x \end{bmatrix} = \begin{bmatrix} 0 & 1 \\ -t & 0 \end{bmatrix}\begin{bmatrix} y \\ x \end{bmatrix}, \quad y(0) = c_1, \quad x(0) = c_2$$

The solution is obtained as

$$\begin{bmatrix} y(t) \\ x(t) \end{bmatrix} = M(t)\begin{bmatrix} c_1 \\ c_2 \end{bmatrix} = \left(I + \int_0^t \begin{bmatrix} 0 & 1 \\ -\tau & 0 \end{bmatrix} d\tau + \right.$$

$$\left. + \int_0^t \begin{bmatrix} 0 & 1 \\ -\tau & 0 \end{bmatrix} \int_0^\tau \begin{bmatrix} 0 & 1 \\ s & 0 \end{bmatrix} ds\, d\tau + \cdots \right)\begin{bmatrix} c_1 \\ c_2 \end{bmatrix}$$

$$= \begin{bmatrix} 1 - \dfrac{t^3}{6} + \cdots & t - \dfrac{t^4}{12} + \cdots \\ \dfrac{-t^2}{2} + \cdots & 1 - \dfrac{t^3}{3} + \cdots \end{bmatrix}\begin{bmatrix} c_1 \\ c_2 \end{bmatrix}$$

giving

$$y(t) = c_1\left(1 - \frac{t^3}{6} + \cdots\right) + c_2\left(t - \frac{t^4}{12} + \cdots\right)$$

which is the sum of the series expansion of Bessel functions of order $-\frac{1}{3}, \frac{1}{3}$, each of them being multiplied by $\sqrt{t}$.

Exercises 5.7

1. Solve the differential equations

$$5\frac{dx}{dt} - 2\frac{dy}{dt} - 4\frac{dz}{dt} = x$$

$$-2\frac{dx}{dt} + 2\frac{dy}{dt} + 2\frac{dz}{dt} = y; \quad x(0) = 1, \quad y(0) = -2, \quad z(0) = 1$$

$$-4\frac{dx}{dt} + 2\frac{dy}{dt} + 5\frac{dz}{dt} = z.$$

Hint: do not invert the matrix of coefficients, since its inverse has inverse eigenvalues and the same eigenvectors.

2. Solve the differential equations

$$\frac{dx}{dt} = -17x + 25z - 1$$

$$\frac{dy}{dt} = 3y + t; \qquad x(0) = 1, \quad y(0) = 2, \quad z(0) = 1$$

$$\frac{dz}{dt} = -9x + 13z + 2$$

3. Show that the solution of the differential equation $dx/dt = Ax + b$, where b is a constant vector and A is nonsingular, is given by $x(t) = e^{At}x(0) + A^{-1}e^{At}b - A^{-1}b$. Show also that if A has eigenvalues with negative real parts, then $\lim_{t\to\infty} x(t) = -A^{-1}b$.

Apply this result to obtain the steady-state solution $i_L(\infty)$, $v_C(\infty)$ of the circuit

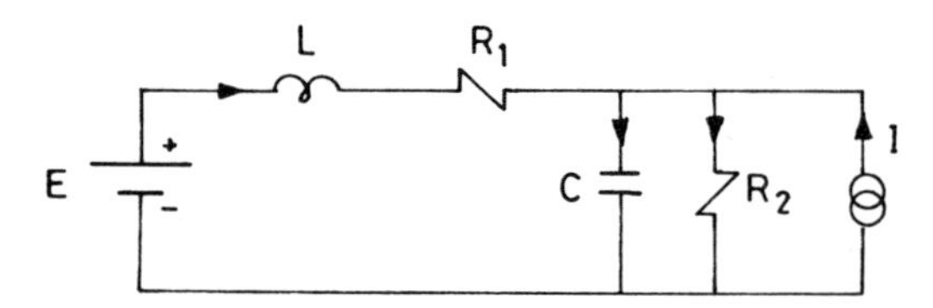

whose state equations are given in Section 5.7 (first Example).

4. Solve the differential equations

$$\frac{d}{dt}\begin{bmatrix} 1 & 1 \\ -2 & -2 \end{bmatrix}\begin{bmatrix} x \\ y \end{bmatrix} + \begin{bmatrix} 1 & 2 \\ -1 & 3 \end{bmatrix}\begin{bmatrix} x \\ y \end{bmatrix} = \begin{bmatrix} e^t \\ t \end{bmatrix}, \quad y(0) = 1$$

Can we choose $x(0)$ arbitrarily?

5. If in the differential equation $dx/dt = Ax + u(t)$, $\| u(\tau) \| < m$, $\forall \tau \in [0, t]$, show that $\| x(t) \| < m_1$ in the same interval.

6. A system is called asymptotically stable in the sense of Lyapunov if the rate of change of the energy storage is negative. Show that, if the energy storage is equal to $E = \frac{1}{2}\langle x, Lx \rangle$, with L being positive definite and $dx/dt = Ax$, the system is stable provided that $LA + A^TL$ is negative definite. Show also that this condition coincides with the property that A has eigenvalues with negative real parts. In stability problems, usually A is given and we solve for L the equation $LA + A^TL = -I$. See Exercise 5.6.19.

7. Show that the solution of the differential equation $dx/dt = Ax + y(t)$, with A semi-simple, is given by

$$x(t) = \sum_i e^{\lambda_i t}E_i x(0) + \sum_i \int_0^t e^{\lambda_i(t-\tau)}E_i y(\tau)\, d\tau$$

where E_i is an eigenprojection of A. Obtain $x(t)$ for a non-semi-simple A.

8. Show that the solution of the differential equation $dx/dt = Ax + y(t)$, with A semi-simple, is given by $x(t) = \sum_i \alpha_i(t)u^i$, where u^i is an eigenvector of A and

$\alpha_i(t)$ is given by

$$\alpha_i(t) = e^{\lambda_i t}\alpha_i(0) + \int_0^t e^{\lambda_i(t-\tau)}\beta_i(\tau)\,d\tau$$

and where $y(t) = \sum_i \beta_i(t)u^i$. This process is called isolation of normal modes, for if $y(t) = 0$, we call the jth mode excited if $x(0)$ is proportional to u^j, i.e. $\alpha_i(0) = 0\ (i \neq j)$. In this case $x(t) = e^{\lambda_j t}x(0)$. For reading about normal modes, refer to Zadeh and Desoer (1963), p. 311. Compare also the expression of $x(t)$ with that obtained in Exercise 7 above.

9. Show that the solution of the differential equation $d^2x/dt^2 + A^2x = 0$, with A being positive definite, is given by

$$x(t) = (\sin At)u + (\cos At)v$$

where u and v are arbitrary vectors. Hence solve

$$\frac{d^2x}{dt^2} + 13x + 5y = 0, \quad \frac{d^2y}{dt^2} + 5x + 13y = 0, \quad x(0) = y(0) = 1.$$

Can you write by intuition the solution of the differential equation $d^2x/dt^2 - A^2x = 0$, with A positive definite?

10. Solve the differential equation $M\,d^2x/dt^2 + Kx = 0$, of a mass-spring system, with M and K being both Hermitian positive definite. *Hint*: use simultaneous diagonalization of M and K (Exercise 4.10.21) by making the change of variables $x(t) = Cy(t)$ and pre-multiplying by C^*. This method is better than pre-multiplying the equations by M^{-1}, since it preserves the properties of the system.

11. For the mass-spring system shown in the figure below whose dynamical

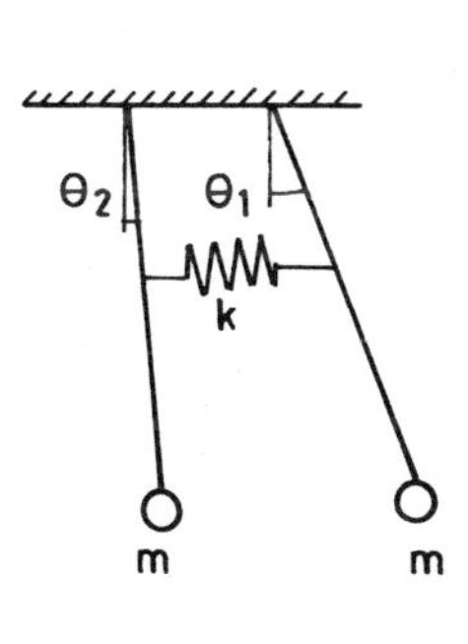

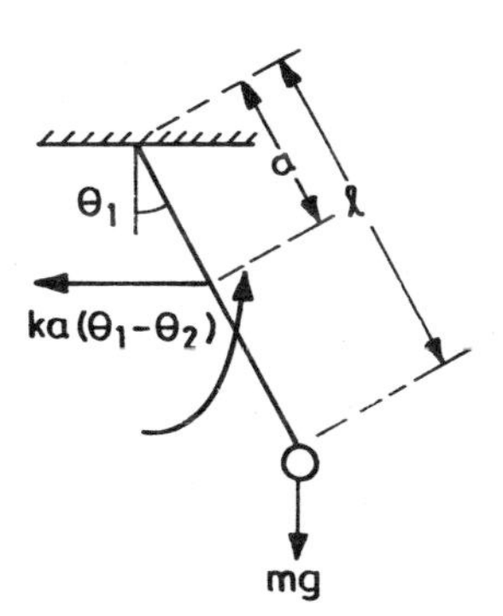

equations are for small oscillations

$$ml^2\ddot{\theta}_1 = -mgl\theta_1 - ka^2(\theta_1 - \theta_2)$$
$$ml^2\ddot{\theta}_2 = -mgl\theta_2 + ka^2(\theta_1 - \theta_2),$$

determine the normal mode vibrations.

12. Show that the solution of the differential equation

$$M\frac{d^2x}{dt^2} + B\frac{dx}{dt} + Kx = f(t)$$

of a system having distinct natural frequencies is given by

$$x(t) = \sum_i c_i e^{\lambda_i t} u^i + (MD^2 + BD + K)^a \sum_i \frac{e^{\lambda_i t}}{\Delta_i} \int_0^t e^{-\lambda_i \tau} f(\tau)\, d\tau$$

where λ_i is a distinct root of $|M\lambda^2 + B\lambda + K| = 0$, u^i is a solution of

$$(M\lambda_i^2 + B\lambda_i + K)u^i = 0, \quad \Delta_i = \lim_{\lambda \to \lambda_i} \frac{|M\lambda^2 + B\lambda + K|}{\lambda - \lambda_i}$$

and

$$D = \frac{d}{dt}$$

For a damped system of many degrees of freedom, the necessary computational effort made in calculating $x(t)$ in this way is enormous, and in practice it is rarely attempted. Instead, the situation becomes simpler by uncoupling the differential equation to isolate the vibration modes. The method is via simultaneous diagonalization. Substituting $x(t) = Ny(t)$, where

$$N^*MN = I, \quad N^*KN = \text{diag}(w_1^2, w_2^2, \ldots, w_n^2)$$

we obtain for the ith mode

$$\ddot{y}_i + c_i \dot{y}_i + w_i^2 y_i = 0, \quad i = 1, \ldots, n$$

if $N^*BN = \text{diag}(c_1, c_2, \ldots, c_n)$. Although N^*BN is not diagonal in general, we often encounter cases where N^*BN is diagonal dominant. So if the natural frequencies are well apart, it is safe to disregard the off-diagonal elements. Discuss the case when M is positive semidefinite. For further reading refer to Pestel and Leckie (1963) and Thomson (1972).

13. Show that the matrizant $M(t) = \exp(\int A\, dt)$. Hence show that $|M(t)| \neq 0$.
14. Solve the differential equation $\ddot{y} + 2t\dot{y} + y = \sin t$, $y(0) = 1$, $\dot{y}(0) = -1$.
15. Solve the differential equation

$$\frac{d}{dt}\begin{bmatrix} x \\ y \end{bmatrix} = \begin{bmatrix} 1 & t \\ -t & 1 \end{bmatrix}\begin{bmatrix} x \\ y \end{bmatrix} + \begin{bmatrix} \sin t \\ \cos t \end{bmatrix}, \quad x(0) = 1, \quad y(0) = 1$$

16. If in the equation $dx/dt = A(t)x$, $A(t)$ is periodic i.e. $A(t + \tau) = A(t)$, show that $x(n\tau) = M^n(\tau)x(0)$.
17. Solve the differential equations

$$\frac{d}{dt}\begin{bmatrix} x \\ y \end{bmatrix} = \begin{bmatrix} \cos t & \sin t \\ -\sin t & \cos t \end{bmatrix}\begin{bmatrix} x \\ y \end{bmatrix}, \quad x(0) = 1, \quad y(0) = -1.$$

18. If $x_1(t), x_2(t), \ldots, x_n(t)$ are solutions of the differential equation

$$\frac{d^n x}{dt^n} + a_{n-1}(t)\frac{d^{n-1}x}{dt^{n-1}} + a_{n-2}(t)\frac{d^{n-2}x}{dt^{n-2}} + \cdots + a_1(t)\frac{dx}{dt} + a_0(t) = 0$$

Show that $x_1(t), x_2(t), \ldots, x_n(t)$ are linearly independent $\forall \tau \in [0, t]$ if

$$\textit{Wronskian} \begin{vmatrix} x_1 & x_2 & \cdots & x_n \\ \dot{x}_1 & \dot{x}_2 & \cdots & \dot{x}_n \\ x_1^{(n-1)} & & & x_n^{(n-1)} \end{vmatrix} \neq 0$$

Discuss its relation with the matrizant of the reduced first-order equation and show that the above condition is equivalent to $|M(t)| \neq 0$.

19. Consider a system of first-order nonlinear differential equations

$$\frac{dx_1}{dt} = f_1(x_1, \ldots, x_n, t)$$

$$\frac{dx_2}{dt} = f_2(x_1, \ldots, x_n, t)$$

$$\vdots$$

$$\frac{dx_n}{dt} = f_n(x_1, \ldots, x_n, t) \quad \text{given } x_1(0), x_2(0), \ldots, x_n(0)$$

Show that the solution for second-order approximation is given by Taylor's series as

$$x(t) \approx x(0) + tF(x(0), 0) + \frac{t^2}{2!} Z(x(0), 0)\, F_1(x(0), 0)$$

where

$$x = \begin{bmatrix} x_1 \\ x_2 \\ \vdots \\ x_n \end{bmatrix}, \quad F = \begin{bmatrix} f_1(x, t) \\ f_2(x, t) \\ \vdots \\ f_n(x, t) \end{bmatrix}, \quad Z = \begin{bmatrix} \dfrac{\partial f_1}{\partial x_1} & \dfrac{\partial f_1}{\partial x_2} & \cdots & \dfrac{\partial f_1}{\partial x_n} & \dfrac{\partial f_1}{\partial t} \\ \vdots & \vdots & & \vdots & \vdots \\ \dfrac{\partial f_n}{\partial x_1} & \dfrac{\partial f_n}{x_2} & \cdots & \dfrac{\partial f_n}{\partial x_n} & \dfrac{\partial f_n}{\partial t} \end{bmatrix},$$

$$F_1 = \begin{bmatrix} F \\ --- \\ 1 \end{bmatrix}$$

There are various numerical methods for solving nonlinear differential equations. The most famous are due to Euler and to Runge and Kutta. In Euler's method we truncate $x(t)$ after the first derivatives to read $x(h) \approx x(0) + hf(x(0), 0)$, where h is small. Euler's method is not usually accurate as the accuracy is a function of the step size. The Runge–Kutta method gives accurate results without a large number of steps, i.e. without the need to make the step size too small. It relies upon calculating the integral in

$$x(t) = x(0) + \int_0^t f(x, \tau)\, dt$$

using Simpson's rule. For further reading about this subject refer to Derrick and Grossman (1976), Chapter 8.

20. Solve the differential equation for $\theta(t)$ representing the motion of a pendulum:

$$\frac{d\theta}{dt} = x, \quad \frac{dx}{dt} = -4.6499 \sin\theta, \quad x(0) = 0, \quad \theta(0) = 2.0944$$

21. If the function $F(x, t)$ is continuous at all points (x, t) in some domain R, $[R/\| x - x^0 \| < a, | t - t_0 | < b]$, and $F(x, t)$ is bounded in R, say $\| F(x, t) \| < m$,

$\forall x, t \in R$, then show that the initial value problem $dx/dt = F(x, t), x(t_0) = x^0$, has at least one solution $x(t)$ which is defined at least for all t in the interval $|t - t_0| < b$.

Hint: write

$$x(t) = x^0 + \int_{t_0}^{t} F(x, \tau)\, d\tau$$

and then take norms.

Show also that the above condition is only a sufficient one, since for example $dy/dx = y, y(0) = 1$ has a solution in $R[R/x \in [0, \sqrt{x}]]$ although $F(y, x)$ is unbounded in R. Find the necessary condition.

22. If both the functions $F(x, t)$ and $\partial F/\partial x$ are continuous for all points (x, t) in some domain R and bounded in R, say

$$\| F(x, t) \| < m, \qquad \left\| \frac{\partial F}{\partial x} \right\| < k$$

show that the differential equation

$$\frac{dx}{dt} = F(x, t), \quad x(0) = x^0$$

has only one solution in R which can be obtained by iteration. *Hint*: write

$$x(t) = x(0) + \int_0^t F(x, \tau)\, d\tau, \quad y(t) = y(0) + \int_0^t F(y, \tau)\, d\tau, \quad y(0) = x(0)$$

and subtract; then take norms and by iteration

$$\| x_n(t) - y_n(t) \| \leqslant m_1 \frac{\left(\int_0^t \left\| \frac{\partial F}{\partial x} \right\| d\tau \right)^{n+1}}{(n+1)!}$$

23. Show that if $\| F(x, y) \| < m$, and $\| F(x, y) - F(x, y') \| \leqslant a \| y - y' \|$, the differential equation $dy/dx = F(x, y), y(0) = b$ has a unique solution.

24. Show that the differential equation $dy/dx = \sqrt{y^2 - 1}, y(0) = 1$ has a solution, but it is not unique.

25. Show that the differential equation $dy/dx = 1/\sqrt{x^2 - 1}, y(1) = 0$ has a unique solution, given by $y = \cosh^{-1} x$.

26. Show that if $\| A(\tau) \| < m, \forall \tau \in [0, t]$. then the differential equation $dx/dt = A(t)x$ has a unique solution for $x(t)$.

27. The linear time-invariant differential equation $dx/dt = Ax + Bu(t)$ is said to be completely controllable if for any initial state x^0 and final state x^1 there exists a finite time $t_1 > 0$ and a real input u defined on $[0, t_1]$ such that

$$x^1 = e^{At_1} x^0 + e^{At_1} \int_0^{t_1} e^{-At} Bu(t)\, dt$$

Show that the system is completely controllable if and only if either or both of the two conditions are satisfied:

(i) The rows of $e^{-At}B$ are linearly independent on $[0, t_1]$.
(ii) The $(n, n \times m)$ controllability matrix $[B \mid AB \mid \cdots \mid A^{n-1}B]$ has rank n. *Hint*: use the Cayley–Hamilton theorem to represent e^{-At}. When should one use the minimal polynomial?

28. The linear time-invariant system represented by $dx/dt = Ax + Bu(t)$, $y(t) = Cx(t) + Du(t)$, where C and D are respectively the output matrix and direct transmission matrix, is said to be *completely observable* if for every initial state x^0, with input $u = 0$, there exists a finite time $t_1 \geqslant 0$, such that the knowledge of the output $y(t)$ over the interval $[0, t_1]$ suffices to determine the initial state x^0. Show that the system is completely observable if and only if the $(n, n \times k)$ observability matrix $[C^* \mid A^*C^* \mid \cdots \mid A^{*n-1}C^*]$ has rank n. For further reading about controllability and observability and other related subjects, refer to Zadeh and Desoer (1963), p. 495, and to Zadeh and Polak (1969), p. 244.

29. For the circuit

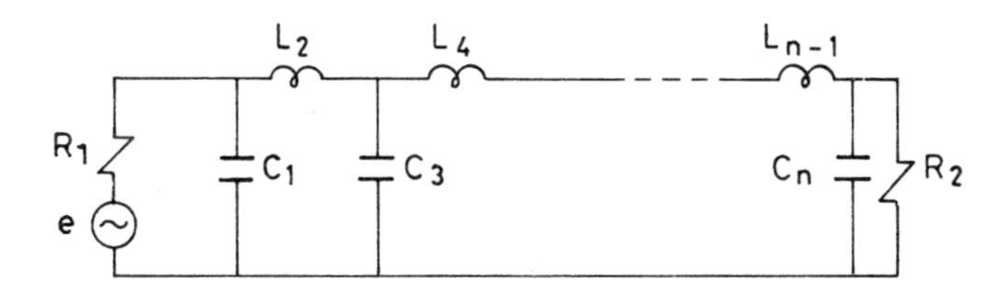

show that, if the state vector x is given by

$$x^T = [v_{C_1}, i_{L_2}, v_{C_3}, i_{L_4}, \ldots, i_{L_{n-1}}, v_{C_n}]$$

the system state matrix is the tridiagonal matrix

$$A = \begin{bmatrix} -\dfrac{1}{R_1C_1} & -\dfrac{1}{C_1} & & \\ \dfrac{1}{L_2} & 0 & -\dfrac{1}{L_2} & \\ & \ddots & \ddots & \ddots \\ & & \dfrac{1}{C_n} & -\dfrac{1}{R_2C_n} \end{bmatrix}$$

Obtain the transfer function (v_{C_n}/e) in the form $c(\lambda I - A)^{-1}b = c(\lambda I - A)^a b / |\lambda I - A|$, where $c = [0, 0, \ldots, 0, 1]$, $b^T = [1/R_1C_1, 0, \ldots, 0]$. Show that $c(\lambda I - A)^a b$ is independent of λ. Show also, using Exercise 4.10.6, that the system natural frequencies (poles of the transfer function or eigenvalues of A) have negative real parts assuring stability if L, C and R have positive values. The characteristic polynomial of A is useful in circuit design; see Temes and Calahan (1967).

CHAPTER SIX

PERTURBATION THEORY

6.1. Introduction

Perturbation theory studies the behaviour of a system subjected to small perturbations in its variables. If the system is represented by a set of linear simultaneous equations of the form $Ax = b$, the problem becomes that of determining x when A exhibits a perturbation of the form $A + \epsilon A_1$, and b a perturbation of the form $b + \epsilon b_1$. The constant ϵ is a scalar quantity much less than unity and is called the perturbation factor. Sometimes ϵ is included in A_1 to mean that A_1 contains small elements relative to those of A.

On the other hand, if the system is represented by the differential equation $\mathrm{d}x/\mathrm{d}t = Ax + y(t)$, the problem becomes that of determining the solution $x(t)$ if A exhibits a perturbation of the form $A + \epsilon A_1$.

It should be stressed that perturbation analysis starts only after we have already obtained the solution of the original system; which means that the theory is there only to explore the change in the behaviour of the system when perturbations take place. Or more precisely the technique used obtains only the deviation in the solution from that of the original system when the latter is already known. To explain the matter by a very simple example, consider the equation

$$ax = b$$

With $a, b \neq 0$, the solution is given by

$$x = \frac{b}{a}$$

Now if a changes to $a + \epsilon$, $|\epsilon| < a$, how does this affect x? We can easily obtain

$$\begin{aligned}\tilde{x} &= \frac{b}{a + \epsilon} \\ &= \frac{b}{a}\left(1 + \frac{\epsilon}{a}\right)^{-1} \\ &= \frac{b}{a}\left(1 - \frac{\epsilon}{a} + \frac{\epsilon^2}{a^2} \cdots\right) \approx x - \frac{\epsilon b}{a^2}\end{aligned}$$

Therefore the deviation in x is approximately equal to $-\epsilon b/a^2$ for first-order perturbations. Note also that this term is exactly equal to $\epsilon\, \mathrm{d}x/\mathrm{d}a$.

6.2. Perturbation of linear equations

Let the solution of the system of linear simultaneous equations

$$Ax = b$$

be known, which is $A^{-1}b$. Then it is required to determine x when A becomes $A + \epsilon A_1$ and b becomes $b + \epsilon b_1$. The technique we shall employ does not compute the whole of x_{new}; instead it will only determine the deviation $x_{new} - x$. The new system of linear simultaneous equations will then read

$$(A + \epsilon A_1)x_{new} = b + \epsilon b_1$$

where x_{new} can easily be shown to be expandable into a convergent series in ϵ, namely

$$x_{new} = x + \epsilon x_{(1)} + \epsilon^2 x_{(2)} + \cdots$$

Therefore substituting in the perturbed system we obtain

$$(A + \epsilon A_1)(x + \epsilon x_{(1)} + \epsilon^2 x_{(2)} + \cdots) = b + \epsilon b_1$$

And grouping terms in ϵ, we get:

$$Ax_{(1)} + A_1 x = b_1$$
$$Ax_{(2)} + A_1 x_{(1)} = 0$$
$$\vdots$$
$$Ax_{(n+1)} + A_1 x_{(n)} = 0$$
$$\vdots$$

From which we deduce that

$$x_{(1)} = -A^{-1}(A_1 x - b_1)$$
$$x_{(2)} = -A^{-1}A_1 x_{(1)}$$
$$\vdots$$
$$x_{(n+1)} = -A^{-1}A_1 x_{(n)}$$
$$\vdots$$

Note that A^{-1} is already known since we used it in calculating x. Therefore the method is less troublesome than inverting $A + \epsilon A_1$; moreover we may only need to know $x_{(1)}$. And if we are interested in having an idea about $\| x_{(1)} \| / \| x \|$ so that we can check the accuracy, or in other words we want to know the error in computing x if there is an error ϵA_1 in measuring the matrix A, we can proceed as follows:

$$Ax_{(1)} = b_1 - A_1 x$$

Hence

$$\| Ax_{(1)} \| = \| b_1 - A_1 x \|$$

However

$$\| Ax_{(1)} \| \geqslant \| A^{-1} \|^{-1} \| x_{(1)} \|$$

and

$$\| b_1 - A_1 x \| \leqslant \| b_1 \| + \| A_1 \| \| x \|$$

Therefore

$$\| A^{-1} \|^{-1} \| x_{(1)} \| \leqslant \| b_1 \| + \| A_1 \| \| x \|$$

from which we obtain

$$\frac{\| x_{(1)} \|}{\| x \|} \leqslant \| A^{-1} \| \left(\frac{\| b_1 \|}{\| x \|} + \| A_1 \| \right)$$

$$= \| A^{-1} \| \| A \| \left(\frac{\| b_1 \|}{\| A \| \| x \|} + \frac{\| A_1 \|}{\| A \|} \right)$$

However

$$\| b \| = \| Ax \| \leqslant \| A \| \| x \|$$

and therefore

$$\frac{\| x_{(1)} \|}{\| x \|} \leqslant \gamma(A) \left(\frac{\| b_1 \|}{\| b \|} + \frac{\| A_1 \|}{\| A \|} \right)$$

Exercises 6.2

1. If $Ax = b$, with $A = \begin{bmatrix} 1 & 2 \\ 3 & 4 \end{bmatrix}$, $b = \begin{bmatrix} 5 \\ 6 \end{bmatrix}$, and if A and b exhibits a change such that $A_{new} = A + \epsilon A_1$, $b_{new} = b + \epsilon b_1$, where

$$A_1 = \begin{bmatrix} 0.1 & 0 \\ 0.02 & 0.3 \end{bmatrix}, \quad b_1 = \begin{bmatrix} 0.01 \\ -0.1 \end{bmatrix}$$

find the new x approximately and calculate the error $\dfrac{\| x_{(1)} \|}{\| x \|}$

2. If, in the equation $(A + \delta A)(x + \delta x) = b + \delta b$, the term $\delta A \delta x$ is not neglected i.e. $A\delta x + \delta A x + \delta A \delta x = \delta b$, show that

$$\frac{\| \delta x \|}{\| x \|} \leqslant \frac{\gamma(A)}{1 - \gamma(A) \left(\dfrac{\| \delta A \|}{\| A \|} \right)} \left(\frac{\| \delta A \|}{\| A \|} + \frac{\| \delta b \|}{\| b \|} \right)$$

3. Show that, if $\| \epsilon A^{-1} A_1 \| < 1$, then $(A + \epsilon A_1)^{-1} = A^{-1} - \epsilon A^{-1} A_1 A^{-1} + \epsilon^2 A^{-1} A_1 A^{-1} A_1 A^{-1} + \cdots$. Hence obtain x, if $(A + \epsilon A_1) x = b + \epsilon b_1$, in a series of ϵ.

4. If $x = (A^T A)^{-1} A^T b$ minimizes $\| Ax - b \|_2$, show that

$$x_{(1)} = (A^T A)^{-1}(A_1^T b + A^T b_1 - A^T A_1 x - A_1^T A x)$$

minimizes approximately $\| (A + \epsilon A_1)(x + \epsilon x_{(1)}) - (b + \epsilon b_1) \|_2$. Obtain also $\| x_{(1)} \| / \| x \|$.

5. If X solves $AX + XB = C$, and $X + \delta X$ solves $(A + \delta A)(X + \delta X) + (X + \delta X)(B + \delta B) = C + \delta C$, show, neglecting second-order terms, that

$$\frac{\| \delta X \|}{\| X \|} \leqslant (\| A \| + \| B \|) \frac{\mathscr{K}(A)\mathscr{K}(B)}{\min\limits_{i,j} | \lambda_i + \mu_j |} \left(\frac{\| \delta C \|}{\| C \|} + \frac{\| \delta A \| + \| \delta B \|}{\| A \| + \| B \|} \right)$$

where λ and μ are the eigenvalues of A and B and $\mathscr{K}(A)$ and $\mathscr{K}(B)$ are respectively their spectral condition numbers. *Hint*: make use of Exercise 5.6.20.

6.3. Perturbations in the eigenvalues and eigenvectors of a matrix

The problem treated in this section is that of calculating the shifts occurring in the eigenvalues and eigenvectors of a matrix A. If A has the eigenvalues $\lambda_1, \lambda_2, \ldots, \lambda_n$ and eigenvectors $u^1, u^2, \ldots, u^n$, then the eigenvalues and eigenvectors of $A + \epsilon A_1$ are the topic we are now investigating. If the eigenvalue problem for A is

$$Au = \lambda u,$$

then the eigenvalue problem for the perturbed system will be

$$(A + \epsilon A_1)\tilde{u} = \tilde{\lambda}\tilde{u}.$$

By a similar method to what we have used for the system of equations, we first show that $\tilde{\lambda}$ is a series convergent in ϵ. From the above equations the vector $\tilde{u}$ has a non-trivial solution only if

$$| \tilde{\lambda} I - A - \epsilon A_1 | = 0$$

However the above determinant can obviously be written in powers of ϵ as follows:

$$| \tilde{\lambda} I - A - \epsilon A_1 | = D_0(\tilde{\lambda}) + \epsilon D_1(\tilde{\lambda}) + \cdots + \epsilon^n D_n(\tilde{\lambda})$$

where

$$D_0(\tilde{\lambda}) = | \tilde{\lambda} I - A |$$

Expanding $| \tilde{\lambda} I - A - \epsilon A_1 |$ in Taylor's series around λ, which is the eigenvalue of the unperturbed matrix, we obtain

$$\begin{aligned} 0 = {} & D_0(\lambda) + \epsilon D_1(\lambda) + \cdots + \epsilon^n D_n(\lambda) \\ & + (\tilde{\lambda} - \lambda)[D_0'(\lambda) + \epsilon D_1'(\lambda) + \cdots] \\ & \vdots \\ & + \frac{(\tilde{\lambda} - \lambda)^m}{m!} [D_0^{(m)}(\lambda) + \epsilon D_1^{(m)}(\lambda) + \cdots] \\ & \vdots \\ & + \frac{(\tilde{\lambda} - \lambda)^n}{n!} D_0^{(n)}(\lambda) \end{aligned}$$

with $D_0(\lambda) = 0$. Now we stop to consider the two cases:

(i) λ is a distinct eigenvalue of A.
(ii) λ is a multiple eigenvalue of A of multiplicity m.

For the first case, the reader can show that the above Taylor's expansion yields for small $(\tilde{\lambda} - \lambda)$:

$$\lambda = \lambda + \epsilon\lambda^{(1)} + \epsilon^2\lambda^{(2)} + \cdots$$

and the problem becomes that of calculating $\lambda^{(1)}, \lambda^{(2)}, \ldots$. The reader can prove that $\tilde{u}$ can also be written in the form

$$\tilde{u} = u + \epsilon u_{(1)} + \epsilon^2 u_{(2)} + \cdots$$

by considering the set of linear simultaneous equations

$$(A + \epsilon A_1 - (\lambda + \epsilon\lambda^{(1)} + \epsilon^2\lambda^{(2)} + \cdots)I)\,\tilde{u} = 0.$$

To compute the shifts in the eigenvalues and eigenvectors we substitute $\tilde{\lambda}_i$ and $\tilde{u}_i$ in the eigenvalue problem of the perturbed system to give

$$(A + \epsilon A_1)(u^i + \epsilon u^i_{(1)} + \cdots) = (\lambda_i + \epsilon\lambda_i^{(1)} + \cdots)(u^i + \epsilon u^i_{(1)} + \cdots)$$

Grouping terms in ϵ, we obtain

$$Au^i_{(1)} + A_1 u^i = \lambda_i u^i_{(1)} + \lambda_i^{(1)} u^i$$

$$Au^i_{(2)} + A_1 u^i_{(1)} = \lambda_i u^i_{(2)} + \lambda_i^{(1)} u^i_{(1)} + \lambda_i^{(2)} u^i$$

$$\vdots$$

$$Au^i_{(n+1)} + A_1 u^i_{(n)} = \lambda_i u^i_{(n+1)} + \lambda_i^{(1)} u^i_{(n)} + \cdots + \lambda_i^{(n+1)} u^i$$

$$\vdots$$

We shall compute only $\lambda_i^{(1)}$ and $u^i_{(1)}$ here; as an exercise the reader can work out a method to compute $\lambda_i^{(2)}, \lambda_i^{(3)}, \ldots, u^i_{(2)}, u^i_{(3)} \ldots$. Taking the inner product of the first equation with v^i, which is the ith eigenvector of A^* i.e. the reciprocal of u^i, we obtain

$$\langle v^i, Au^{(i)}_{(1)}\rangle + \langle v^i, A_1 u^i\rangle = \lambda_i\langle v^i, u^i_{(1)}\rangle + \lambda_i^{(1)}$$

However

$$\langle v^i, Au^i_{(1)}\rangle = \lambda_i\langle v^i, u^i_{(1)}\rangle.$$

Hence we obtain

$$\lambda_i^{(1)} = \langle v^i, A_1 u^i\rangle$$

while taking the inner product with v^k, $k \neq i$, we obtain

$$\langle v^k, Au^i_{(1)}\rangle + \langle v^k, A_1 u^i\rangle = \lambda_i\langle v^k, u^i_{(1)}\rangle.$$

To obtain $u^i_{(1)}$ we write

$$u^i_{(1)} = \sum_{j=1}^{n} c_{ji} u^j$$

The above expansion is valid if A is a semi-simple matrix, since the eigenvectors are linearly independent, and upon substituting for $u^i_{(1)}$ we obtain

$$\left\langle v^k, A \sum_{j=1}^{n} c_{ji} u^j \right\rangle + \langle v^k, A_1 u^i \rangle = \lambda_i \left\langle v^k, \sum_{j=1}^{n} c_{ji} u^j \right\rangle$$

i.e.

$$\lambda_k c_{ki} + \langle v^k, A_1 u^i \rangle = \lambda_i c_{ki}$$

and we directly obtain

$$c_{ki} = \frac{\langle v^k, A_1 u^i \rangle}{\lambda_i - \lambda_k}$$

i.e.

$$u^i_{(1)} = \sum_{\substack{k=1 \\ k \neq i}}^{n} \frac{\langle v^k, A_1 u^i \rangle}{\lambda_i - \lambda_k} u^k$$

Therefore we have established

THEOREM 1. If λ_i is a distinct eigenvalue of a semi-simple matrix A with corresponding eigenvector u^i, the eigenvalue $\tilde{\lambda}_i$ and corresponding eigenvector $\tilde{u}^i$ of the perturbed matrix $A + \epsilon A_1$ are given for first-order approximation by

$$\tilde{\lambda}_i \approx \lambda_i + \epsilon \langle v^i, A_1 u^i \rangle$$

and

$$\tilde{u}^i \approx u^i + \epsilon \sum_{k=1, k \neq i}^{n} \frac{\langle v^k, A_1 u^i \rangle}{\lambda_i - \lambda_k} u^k$$

where $u^1, u^2, \ldots, u^n$ are the eigenvectors of A and $v^1, v^2, \ldots, v^n$ their reciprocal basis.

Example. Let

$$A = \begin{bmatrix} 6 & 8 \\ 8 & -6 \end{bmatrix}, \quad A_1 = \begin{bmatrix} 2 & -1 \\ 0 & 2 \end{bmatrix}$$

The eigenvalues of A are $10, -10$

The eigenvectors of A are:

$$\begin{bmatrix} \frac{2}{\sqrt{5}} \\ \frac{1}{\sqrt{5}} \end{bmatrix}, \quad \begin{bmatrix} \frac{1}{\sqrt{5}} \\ \frac{-2}{\sqrt{5}} \end{bmatrix}$$

Their reciprocal bases are:

$$\begin{bmatrix} \frac{2}{\sqrt{5}} \\ \frac{1}{\sqrt{5}} \end{bmatrix}, \quad \begin{bmatrix} \frac{1}{\sqrt{5}} \\ \frac{-2}{\sqrt{5}} \end{bmatrix}$$

The eigenvalues of $A + \epsilon A_1$ are the roots of

$$\begin{vmatrix} 6 + 2\epsilon - \lambda & 8 - \epsilon \\ 8 & -6 + 2\epsilon - \lambda \end{vmatrix} = 0$$

giving

$$\tilde{\lambda}_1 = 2\epsilon + 10\sqrt{1 - \frac{32\epsilon}{400}}, \quad \tilde{\lambda}_2 = 2\epsilon - 10\sqrt{1 - \frac{32\epsilon}{400}}$$

and their eigenvectors are obtained by solving $(\tilde{\lambda} I - A - \epsilon A_1)\tilde{u} = 0$, giving

$$\tilde{u}^1 = \begin{bmatrix} \frac{2}{\sqrt{5}} - \frac{\epsilon}{100\sqrt{5}} \\ \frac{1}{\sqrt{5}} + \frac{\epsilon}{50\sqrt{5}} \end{bmatrix}, \quad \tilde{u}^2 = \begin{bmatrix} \frac{1}{\sqrt{5}} - \frac{4\epsilon}{50\sqrt{5}} \\ \frac{-2}{\sqrt{5}} - \frac{2\epsilon}{50\sqrt{5}} \end{bmatrix}$$

Using perturbation techniques we obtain:

$$\tilde{\lambda}_1 \approx 10 + \epsilon \begin{bmatrix} \frac{2}{\sqrt{5}} & \frac{1}{\sqrt{5}} \end{bmatrix} \begin{bmatrix} 2 & -1 \\ 0 & 2 \end{bmatrix} \begin{bmatrix} \frac{2}{\sqrt{5}} \\ \frac{1}{\sqrt{5}} \end{bmatrix} = 10 + \frac{8}{5}\epsilon$$

$$\tilde{\lambda}_2 \approx -10 + \epsilon \begin{bmatrix} \frac{1}{\sqrt{5}} & \frac{-2}{\sqrt{5}} \end{bmatrix} \begin{bmatrix} 2 & -1 \\ 0 & 2 \end{bmatrix} \begin{bmatrix} \frac{1}{\sqrt{5}} \\ \frac{-2}{\sqrt{5}} \end{bmatrix} = -10 + \frac{12}{5}\epsilon$$

$$\tilde{u}^1 \approx u^1 + \frac{\epsilon}{20}\begin{bmatrix} \frac{1}{\sqrt{5}} & \frac{-2}{\sqrt{5}} \end{bmatrix}\begin{bmatrix} 2 & -1 \\ 0 & 2 \end{bmatrix}\begin{bmatrix} \frac{2}{\sqrt{5}} \\ \frac{1}{\sqrt{5}} \end{bmatrix}\begin{bmatrix} \frac{1}{\sqrt{5}} \\ \frac{-2}{\sqrt{5}} \end{bmatrix} = u^1 + \epsilon\begin{bmatrix} \frac{-1}{100\sqrt{5}} \\ \frac{1}{50\sqrt{5}} \end{bmatrix}$$

$$\tilde{u}^2 \approx u^2 + \frac{\epsilon}{-20}\begin{bmatrix} \frac{2}{\sqrt{5}} & \frac{1}{\sqrt{5}} \end{bmatrix}\begin{bmatrix} 2 & -1 \\ 0 & 2 \end{bmatrix}\begin{bmatrix} \frac{1}{\sqrt{5}} \\ \frac{-2}{\sqrt{5}} \end{bmatrix}\begin{bmatrix} \frac{2}{\sqrt{5}} \\ \frac{1}{\sqrt{5}} \end{bmatrix} = u^2 + \epsilon\begin{bmatrix} \frac{-4}{50\sqrt{5}} \\ \frac{-2}{50\sqrt{5}} \end{bmatrix}$$

which are the same results obtained above for small $|\epsilon|$.

The reader will have noticed that the above formulae were deduced on the basis that A is semi-simple but it will be interesting for him to show that they are still the same with minor modifications if A is non-semi-simple.

Now we move to the second case, where λ is a multiple eigenvalue of A with multiplicity equal to m. Going back to Taylor's expansion in $(\tilde{\lambda} - \lambda)$, we notice that not only $D_0(\lambda) = 0$; the reader can show that

$$D_0'(\lambda) = D_0''(\lambda) = \cdots = D_0^{(m-1)}(\lambda) = 0$$

and we obtain for small enough ϵ:

$$\frac{(\tilde{\lambda} - \lambda)^n}{n!} D_0^{(n)}(\lambda) + \frac{(\tilde{\lambda} - \lambda)^{n-1}}{(n-1)!} (D_0^{(n-1)}(\lambda) + \epsilon D_1^{(n-1)}(\lambda)) + \cdots +$$

$$+ \frac{(\tilde{\lambda} - \lambda)^m}{m!} (D_0^{(m)}(\lambda) + \epsilon D_1^{(m)}(\lambda)) + \frac{(\tilde{\lambda} - \lambda)^{m-1}}{(m-1)!} \epsilon D_1^{(m-1)}(\lambda) + \cdots +$$

$$+ (\tilde{\lambda} - \lambda)\epsilon D_1'(\lambda) + \epsilon D_1(\lambda) = 0,$$

where $D_0^{(n)}(\lambda) = n!$ Let the eigenvalues of A be $\lambda, \ldots, \lambda, \lambda_{m+1}, \ldots, \lambda_n$ and the eigenvalues of $A + \epsilon A_1$ be $\tilde{\lambda}_1, \ldots, \tilde{\lambda}_m, \tilde{\lambda}_{m+1}, \ldots, \tilde{\lambda}_n$. Forming the product and sum of the roots of the above equation, we obtain

$$\prod_{i=1}^{n} (\tilde{\lambda}_i - \lambda) = \prod_{i=1}^{m} (\tilde{\lambda}_i - \lambda) \prod_{i=m+1}^{n} (\tilde{\lambda}_i - \lambda) = (-)^n \epsilon D_1(\lambda)$$

and

$$\sum_{i=1}^{n} (\tilde{\lambda}_i - \lambda) = \sum_{i=1}^{m} (\tilde{\lambda}_i - \lambda) + \sum_{i=m+1}^{n} (\tilde{\lambda}_i - \lambda) = \frac{-1}{(n-1)!} [D_0^{(n-1)}(\lambda) + \epsilon D_1^{(n-1)}(\lambda)]$$

Now put

$$\tilde{\lambda}_i - \lambda = (\tilde{\lambda}_i - \lambda_i) + (\lambda_i - \lambda), \quad i = m+1, \ldots, n$$

But as $\lambda_i, i = m + 1, \ldots, n$, is a distinct eigenvalue of A, we obtain

$$\tilde{\lambda}_i - \lambda_i \approx \lambda_i^{(1)}\epsilon, \quad i = m + 1, \ldots, n$$

and the product and sum of the roots become approximately

$$\prod_{i=1}^{m} (\tilde{\lambda}_i - \lambda) \prod_{i=m+1}^{n} (\lambda_i - \lambda) \approx (-)^n \epsilon D_1(\lambda)$$

and

$$\sum_{i=1}^{m} (\tilde{\lambda}_i - \lambda) \approx \frac{-1}{(n-1)!} [D_0^{(n-1)}(\lambda) + \epsilon D_1^{(n-1)}(\lambda)]$$

$$-\epsilon \sum_{i=m+1}^{n} \lambda_i^{(1)} - \sum_{i=m+1}^{n} (\lambda_i - \lambda)$$

Now, from the characteristic equation of the matrix A, the reader can find easily that

$$\prod_{i=m+1}^{n} (\lambda_i - \lambda) = (-)^{n-m} \frac{D_0^{(m)}(\lambda)}{m!}$$

and

$$\frac{1}{(n-1)!} D_0^{(n-1)}(\lambda) + \sum_{i=m+1}^{n} (\lambda_i - \lambda) = 0$$

The equations for the product and sum therefore reduce to

$$\prod_{i=1}^{m} (\tilde{\lambda}_i - \lambda) \approx (-)^m \frac{m! \epsilon D_1(\lambda)}{D_0^{(m)}(\lambda)}$$

and

$$\sum_{i=1}^{m} (\tilde{\lambda}_i - \lambda) \approx -\epsilon \left(\sum_{i=m+1}^{n} \lambda_i^{(1)} + \frac{D_1^{(n-1)}(\lambda)}{(n-1)!} \right)$$

From which we conclude that $\tilde{\lambda}_i, i = 1, \ldots, m$ can be expanded into a series of $\sqrt[m]{\epsilon}$ (Puiseux expansion) as follows:

$$\tilde{\lambda}_i = \lambda + d_i^{(1)} \sqrt[m]{\epsilon} + d_i^{(2)} \sqrt[m]{\epsilon^2} + \cdots + d_i^{(m)} \epsilon + d_i^{(m+1)} \sqrt[m]{\epsilon^{m+1}} + \cdots, \quad i = 1, \ldots, m$$

Therefore for small enough ϵ, $| \tilde{\lambda} I - A - \epsilon A_1 | = 0$ has m roots lying on the circumference of the circle with centre $\lambda + d^{(m)}\epsilon$ and radius $| d_i^{(1)} \sqrt[m]{\epsilon} |$. $d_i^{(1)}$ and $d^{(m)}$ are given by

$$d_i^{(1)} = \sqrt[m]{-\frac{m! D_1(\lambda)}{D_0^{(m)}(\lambda)}} \, e^{2ji\pi/m}, \quad j = \sqrt{-1}, \quad i = 1, \ldots, m$$

and

$$d^{(m)} = -\frac{1}{m}\left(\sum_{i=m+1}^{n} \lambda_i^{(1)} + \frac{D_1^{(n-1)}(\lambda)}{(n-1)!}\right)$$

THEOREM 2. If λ is a semi-simple eigenvalue of A of multiplicity m, the eigenvalue $\tilde{\lambda}$ of the perturbed matrix $A + \epsilon A_1$ is given in the form

$$\tilde{\lambda}_i = \lambda + d_i^{(m)}\epsilon + d_i^{(m+1)}\sqrt[m]{\epsilon^{m+1}} + d_i^{(m+2)}\sqrt[m]{\epsilon^{m+2}} + \cdots, \quad i = 1, \ldots, m$$

Proof. To prove the expansion above is to prove that

$$d_i^{(1)} = d_i^{(2)} \ldots = d_i^{(m-1)} = 0.$$

Going back to Taylor's expansion in $(\tilde{\lambda} - \lambda)$, if one can prove that

$$D_1(\lambda) = D_2(\lambda) = \cdots = D_{m-1}(\lambda) = 0$$

then the Taylor's expansion takes the form

$$\frac{(\tilde{\lambda} - \lambda)^n}{n!} D_0^{(n)}(\lambda) + \frac{(\tilde{\lambda} - \lambda)^{n-1}}{(n-1)!} (D_0^{(n-1)}(\lambda) + \epsilon D_1^{(n-1)}(\lambda)) + \cdots +$$

$$+ \frac{(\tilde{\lambda} - \lambda)^m}{m!} (D_0^{(m)}(\lambda) + \epsilon D_1^{(m)}(\lambda)) + \frac{(\tilde{\lambda} - \lambda)^{m-1}}{(m-1)!} \epsilon D_1^{(m-1)}(\lambda) + \cdots +$$

$$+ (\tilde{\lambda} - \lambda)\epsilon D_1'(\lambda) + \epsilon^m D_m(\lambda) = 0, \quad D_0^{(n)} = n!$$

The product $\prod_{i=1}^{m} (\tilde{\lambda}_i - \lambda)$ is calculated as before and is given by

$$\prod_{i=1}^{m} (\tilde{\lambda}_i - \lambda) \approx (-)^m \frac{m!\epsilon^m D_m(\lambda)}{D_0^{(m)}(\lambda)}$$

which means that $(\tilde{\lambda}_i - \lambda)$ is proportional to ϵ only and not to $\sqrt[m]{\epsilon}$. To show that $D_1(\lambda), \ldots, D_{m-1}(\lambda)$ are all zero if λ is a semi-simple eigenvalue of A of multiplicity m, we use the formula

$$|\mu I - A - \epsilon A_1| = D_0(\mu) + \epsilon D_1(\mu) + \cdots + \epsilon^{m-1} D_{m-1}(\mu) + \cdots + \epsilon^n D_n(\mu)$$

However

$$|\mu I - A - \epsilon A_1| = |\mu I - A|\,|I - \epsilon(\mu I - A)^{-1} A_1|$$

where $|I - \epsilon(\mu I - A)^{-1} A_1|$ is expanded in Taylor's series around $\epsilon = 0$ using Exercise 2.2.13, in the following form

$$|I - \epsilon(\mu I - A)^{-1} A_1| = \exp\left(-\sum_{k=1}^{\infty} (\epsilon^k/k) \operatorname{tr}[(\mu I - A)^{-1} A_1]^k\right)$$

Comparing both expansions we obtain

$$D_0(\mu) = |\mu I - A|$$
$$D_1(\mu) = -|\mu I - A| \operatorname{tr}(\mu I - A)^{-1}A_1$$
$$D_2(\mu) = \tfrac{1}{2}|\mu I - A|((\operatorname{tr}(\mu I - A)^{-1}A_1)^2 - \operatorname{tr}((\mu I - A)^{-1}A_1)^2)$$
$$\vdots$$
$$D_{m-1}(\mu) = 0((\operatorname{tr}(\mu I - A)^{-1}A_1)^{m-1}, (\operatorname{tr}(\mu I - A)^{-1}A_1)\operatorname{tr}((\mu I - A)^{-1}A_1)^{m-2}, \ldots, \operatorname{tr}((\mu I - A)^{-1}A_1)^{m-1})$$

From which we obtain

$$D_0(\mu)|_{\mu=\lambda} = |\lambda I - A| = 0$$
$$D_1(\mu)|_{\mu=\lambda} = -\operatorname{tr}(\mu I - A)^a|_{\mu=\lambda}A_1 = -\operatorname{tr} T^{-1}(\mu I - A)^a|_{\mu=\lambda}TT^{-1}A_1T$$
$$= -\operatorname{tr}\operatorname{diag}\left(\frac{|\mu I - A|}{\mu - \lambda}, \frac{|\mu I - A|}{\mu - \lambda}, \ldots, \frac{|\mu I - A|}{\mu - \lambda}, \frac{|\mu I - A|}{\mu - \lambda_{m+1}}, \ldots, \frac{|\mu I - A|}{\mu - \lambda_n}\right)_{\mu=\lambda} T^{-1}A_1T$$

The term

$$\left.\frac{|\mu I - A|}{\mu - \lambda}\right|_{\mu=\lambda} = \left.\frac{D_0(\mu)}{\mu - \lambda}\right|_{\mu=\lambda} = \frac{0}{0}$$

Differentiating w.r.t. μ and putting $\mu = \lambda$ we obtain

$$\left.\frac{|\mu I - A|}{\mu - \lambda}\right|_{\mu=\lambda} = D_0'(\lambda) = 0$$

Hence

$$D_1(\lambda) = 0$$

$D_2(\mu)$ will contain terms of the form

$$\left.\frac{|\mu I - A|}{(\mu - \lambda)^2}\right|_{\mu=\lambda} = \left.\frac{D_0(\mu)}{(\mu - \lambda)^2}\right|_{\mu=\lambda}$$

and differentiating twice we obtain

$$\left.\frac{|\mu I - A|}{(\mu - \lambda)^2}\right|_{\mu=\lambda} = \frac{D_0''(\lambda)}{2} = 0$$

Hence

$$D_2(\lambda) = 0$$

The same is applied to $D_3(\lambda), \ldots, D_{m-1}(\lambda)$, which are all zero. Of course $D_m(\lambda) \neq 0$ since the latter contains terms of the form

$$\left.\frac{D_0(\mu)}{(\mu-\lambda)^m}\right|_{\mu=\lambda}$$

which is non-zero since $D_0^{(m)}(\lambda) \neq 0$, and the proof is complete.

Now we move to calculate the shift in the eigenvalue λ for small-order perturbations. As the degeneracy is split, the eigenvector $\tilde{u}^i$ will not be identical to u^i when $\epsilon = 0$. Let the eigenvalues of A be $\lambda, \ldots, \lambda, \lambda_{m+1}, \ldots, \lambda_n$ and the corresponding eigenvectors be $u^1, \ldots, u^m, u^{m+1}, \ldots, u^n$. As the matrix A is semi-simple, the eigenvectors form a complete basis, and x^i is also an eigenvector of A if it is expressed in the following form

$$x^i = \sum_{j=1}^{m} c_{ij} u^j$$

The perturbed eigenvector $\tilde{x}^i$ can be expressed in the form

$$\tilde{x}^i = x^i + \epsilon x^i_{(1)} + \sqrt[m]{\epsilon^{m+1}} x^i_{(2)} + \cdots$$

The reader can prove the above representation by considering the set of linear simultaneous equations

$$(A + \epsilon A_1 - (\lambda + \epsilon\lambda_i^{(1)} + \sqrt[m]{\epsilon^{m+1}}\lambda_i^{(2)} + \cdots)I)\,\tilde{x}^i = 0$$

The problem is therefore reduced to determine the values c_{ij} such that the representation of $\tilde{x}^i$ exists. Substituting for $\tilde{\lambda}_i$ and $\tilde{x}^i$ in the eigenvalue problem of the perturbed matrix we obtain

$$(A + \epsilon A_1)(x^i + \epsilon x^i_{(1)} + \cdots) = (\lambda + \epsilon\lambda_i^{(1)} + \cdots)(x^i + \epsilon x^i_{(1)} + \cdots)$$

Grouping terms in ϵ we get

$$Ax^i_{(1)} + A_1 x^i = \lambda x^i_{(1)} + \lambda_i^{(1)} x^i$$

And as the eigenvectors are linearly independent, we can put $x^i_{(1)}$ in the form

$$x^i_{(1)} = \sum_{k=1}^{n} a_{ik} u^k$$

Substituting we obtain

$$A \sum_{k=1}^{n} a_{ik} u^k + A_1 \sum_{j=1}^{m} c_{ij} u^j = \lambda \sum_{k=1}^{n} a_{ik} u^k + \lambda_i^{(1)} \sum_{j=1}^{m} c_{ij} u^j$$

Taking the inner product with $v^1, v^2, \ldots, v^m$, which are the reciprocal vectors of $u^1, u^2, \ldots, u^m$, we obtain

$$\lambda a_{is} + \sum_{j=1}^{m} c_{ij} \langle v^s, A_1 u^j \rangle = \lambda a_{is} + \lambda_i^{(1)} c_{is}, \quad s = 1, \ldots, m$$

giving

$$\begin{bmatrix} \lambda_i^{(1)} - \langle v^1, A_1, u^1 \rangle & -\langle v^1, A_1 u^2 \rangle & \cdots & -\langle v^1, A_1 u^m \rangle \\ -\langle v^2, A_1 u^1 \rangle & \lambda_i^{(1)} - \langle v^2, A_1 u^2 \rangle & \cdots & \\ \vdots & & & \\ -\langle v^m, A_1 u^1 \rangle & & & \lambda_i^{(1)} - \langle v^m, A_1 u^m \rangle \end{bmatrix} \begin{bmatrix} c_{i1} \\ c_{i2} \\ \vdots \\ c_{im} \end{bmatrix} = 0$$

The above equations have a solution for $c_{i1}, c_{i2}, \ldots, c_{im}$ only if the matrix on the left-hand side is singular, i.e. its determinant vanishes. This shows that $\lambda_i^{(1)}$ is an eigenvalue of the matrix S given by

$$S = \begin{bmatrix} v^{1*} \\ v^{2*} \\ \vdots \\ v^{m*} \end{bmatrix} A_1 \begin{bmatrix} u^1 u^2 \ldots u^m \end{bmatrix}$$

while taking the inner product with $v^{m+1}, \ldots, v^n$ we obtain

$$\lambda_s a_{is} + \sum_{j=1}^{m} c_{ij} \langle v^s, A_1 u^j \rangle = \lambda a_{is}, \quad s = m+1, \ldots, n$$

giving

$$a_{is} = \sum_{j=1}^{m} c_{ij} \frac{\langle v^s, A_1 u^j \rangle}{\lambda - \lambda_s}$$

From this we find

$$x_{(1)}^i = \sum_{\substack{k=1 \\ k \neq 1, \ldots, m}}^{n} \frac{\sum_{j=1}^{m} c_{ij} \langle v^k, A_1 u^j \rangle}{\lambda - \lambda_k} u^k$$

Therefore we have established

THEOREM 3. If λ is an eigenvalue of multiplicity m of a semi-simple matrix A, with corresponding eigenvectors $u^1, \ldots, u^m$, the eigenvalues $\tilde{\lambda}_1, \ldots, \tilde{\lambda}_m$ and corresponding eigenvectors $\tilde{x}^1, \ldots, \tilde{x}^m$ of $A + \epsilon A_1$ are given for first-order approximation by

$$\tilde{\lambda}_i \approx \lambda + \epsilon \lambda_i^{(1)}, \quad i = 1, \ldots, m$$

where $\lambda_1^{(1)}, \lambda_2^{(1)}, \ldots, \lambda_m^{(1)}$ are the eigenvalues of

$$S = \begin{bmatrix} v^{1*} \\ \vdots \\ v^{m*} \end{bmatrix} A_1 \begin{bmatrix} u^1 \ldots u^m \end{bmatrix}$$

and

$$\tilde{x}^i \approx x^i + \epsilon \sum_{k=m+1}^{n} \frac{\sum_{j=1}^{m} c_{ij}\langle v^k, A_1 u^j\rangle}{\lambda - \lambda_k} u^k, \quad i = 1, \ldots, m$$

where $u^1, u^2, \ldots, u^n$ are the eigenvectors of A, and $v^1, v^2, \ldots, v^n$ their reciprocal basis. The values of c_{ij} are obtained by solving the set of linear simultaneous equations

$$[\lambda_i^{(1)} I - S]\begin{bmatrix} c_{i1} \\ \vdots \\ c_{im} \end{bmatrix} = 0, \quad i = 1, 2, \ldots, m$$

and

$$x^i = \sum_{j=1}^{m} c_{ij} u^j$$

The reader will have noticed that the above formulae were derived on the assumption that A is semi-simple; therefore he is advised to carry out again the derivation when A is non-semi-simple, while of course λ is a semi-simple eigenvalue. He will find that the formulae have exhibited minor modifications.

One last remark should be said about the perturbation occurring in a semi-simple eigenvalue λ of multiplicity m, namely that if $\lambda_i^{(1)}, i = 1, \ldots, m$ were found to be semi-simple eigenvalues of S, a reduction process can be pursued, and in that case

$$\tilde{\lambda}_i = \lambda + \epsilon\lambda_i^{(1)} + \epsilon^2\lambda_i^{(2)} + \sqrt[m]{\epsilon^{2m+1}}\,\lambda_i^{(3)} + \cdots$$

Of course, the reduction process can continue indefinitely as in the case of real symmetric or Hermitian matrices.

THEOREM 4. If λ is an eigenvalue of multiplicity m of a Hermitian matrix A, with corresponding eigenvectors $u^1, u^2, \ldots, u^m$, the eigenvalues $\tilde{\lambda}_1, \ldots, \tilde{\lambda}_m$ and corresponding eigenvectors $\tilde{u}^1, \ldots, \tilde{u}^m$ of $A + \epsilon A_1$ with A_1 Hermitian are represented in a convergent series in ϵ, i.e.

$$\tilde{\lambda}_i = \lambda + \epsilon\lambda_i^{(1)} + \epsilon^2\lambda_i^{(2)} + \epsilon^3\lambda_i^{(3)} + \cdots$$
$$\tilde{u}^i = u^i + \epsilon u^i_{(1)} + \epsilon^2 u^i_{(2)} + \epsilon^3 u^i_{(3)} + \cdots$$

Proof. The proof follows directly from the Puiseux expansion

$$\tilde{\lambda}_i = \lambda + d_i^{(1)}\sqrt[m]{\epsilon} + d_i^{(2)}\sqrt[m]{\epsilon^2} + \cdots + d_i^{(m)}\epsilon + d_i^{(m+1)}\sqrt[m]{\epsilon^{m+1}} + \cdots$$
$$\tilde{u}^i = u^i + u^i_{(1)}\sqrt[m]{\epsilon} + u^i_{(2)}\sqrt[m]{\epsilon^2} + \cdots + u^i_{(m)}\epsilon + u^i_{(m+1)}\sqrt[m]{\epsilon^{m+1}} + \cdots$$

and noticing that λ and $\tilde{\lambda}_i$ are both real, since A and $A + \epsilon A_1$ are both Hermitian; therefore the imaginary terms $d_i^{(1)}, \ldots, d_i^{(m-1)}, d_i^{(m+1)}, \ldots$ must be zero.

Consequently $u^i_{(1)}, \ldots, u^i_{(m-1)}, u^i_{(m+1)}, \ldots$ must also be zero, as the reader can verify, and the proof is thus complete.

Example. Let

$$A = \begin{bmatrix} 6 & 8 & 0 \\ 8 & -6 & 0 \\ 0 & 0 & 10 \end{bmatrix}, \quad A_1 = \begin{bmatrix} 0 & -1 & 0 \\ -1 & 0 & 0 \\ 0 & 0 & 1 \end{bmatrix}$$

The eigenvalues of A are $\lambda = 10, 10, -10$.
The eigenvectors of A are

$$\begin{bmatrix} \frac{2}{\sqrt{5}} \\ \frac{1}{\sqrt{5}} \\ 0 \end{bmatrix}, \begin{bmatrix} 0 \\ 0 \\ 1 \end{bmatrix}, \begin{bmatrix} \frac{1}{\sqrt{5}} \\ \frac{-2}{\sqrt{5}} \\ 0 \end{bmatrix}$$

The eigenvalues of $A + \epsilon A_1$ are the roots of

$$\begin{vmatrix} 6-\lambda & 8-\epsilon & 0 \\ 8-\epsilon & -6-\lambda & 0 \\ 0 & 0 & 10+\epsilon-\lambda \end{vmatrix} = 0$$

giving

$$\tilde{\lambda}_1 = \sqrt{100+\epsilon^2-16\epsilon}, \quad \tilde{\lambda}_2 = 10+\epsilon, \quad \tilde{\lambda}_3 = -\sqrt{100+\epsilon^2-16\epsilon}$$

and the eigenvectors are, for $\tilde{\lambda}_1$ and $\tilde{\lambda}_2$,

$$\tilde{u}^1 = \begin{bmatrix} 2+\frac{3\epsilon}{20} \\ 1 \\ 0 \end{bmatrix}, \quad \tilde{u}^2 = \begin{bmatrix} 0 \\ 0 \\ 1 \end{bmatrix}$$

Using perturbation techniques, we obtain

$$\begin{bmatrix} v^{1*} \\ v^{2*} \end{bmatrix} A_1 \begin{bmatrix} u^1 & u^2 \end{bmatrix} = \begin{bmatrix} \frac{2}{\sqrt{5}} & \frac{1}{\sqrt{5}} & 0 \\ 0 & 0 & 1 \end{bmatrix} \begin{bmatrix} 0 & -1 & 0 \\ -1 & 0 & 0 \\ 0 & 0 & 1 \end{bmatrix} \begin{bmatrix} \frac{2}{\sqrt{5}} & 0 \\ \frac{1}{\sqrt{5}} & 0 \\ 0 & 1 \end{bmatrix}$$

$$= \begin{bmatrix} \frac{-4}{5} & 0 \\ 0 & 1 \end{bmatrix}$$

hence

$$\tilde{\lambda}_1 \approx 10 - \frac{4\epsilon}{5}, \quad \tilde{\lambda}_2 = 10 + \epsilon$$

and the eigenvectors are

$$\tilde{x}^1 = c_{11}u^1 + c_{12}u^2 + \frac{\epsilon}{20}(c_{11}\langle v^3, A_1 u^1\rangle + c_{12}\langle v^3, A_1 u^2\rangle)u^3$$

$$\tilde{x}^2 = c_{21}u^1 + c_{22}u^2 + \frac{\epsilon}{20}(c_{21}\langle v^3, A_1 u^1\rangle + c_{22}\langle v^3, A_1 u^2\rangle)u^3$$

where c_{11}, c_{12}, c_{21} and c_{22} are obtained from solving

$$\begin{bmatrix} \lambda_i^{(1)} + \frac{4}{5} & 0 \\ 0 & \lambda_i^{(1)} - 1 \end{bmatrix} \begin{bmatrix} c_{i1} \\ c_{i2} \end{bmatrix} = 0$$

Substituting $\lambda_i^{(1)} = -\frac{4}{5}$ once and $\lambda_i^{(1)} = 1$ we obtain

$$\begin{bmatrix} 0 & 0 \\ 0 & -\frac{9}{5} \end{bmatrix} \begin{bmatrix} c_{11} \\ c_{12} \end{bmatrix} = 0 \Rightarrow \begin{bmatrix} c_{11} \\ c_{12} \end{bmatrix} = \begin{bmatrix} 1 \\ 0 \end{bmatrix}$$

$$\begin{bmatrix} \frac{9}{5} & 0 \\ 0 & 0 \end{bmatrix} \begin{bmatrix} c_{21} \\ c_{22} \end{bmatrix} = 0 \Rightarrow \begin{bmatrix} c_{21} \\ c_{22} \end{bmatrix} = \begin{bmatrix} 0 \\ 1 \end{bmatrix}$$

Therefore

$$\tilde{x}^1 \approx \begin{bmatrix} \frac{2}{\sqrt{5}} \\ \frac{1}{\sqrt{5}} \\ 0 \end{bmatrix} + \frac{\epsilon}{20} \begin{bmatrix} \frac{1}{\sqrt{5}} & \frac{-2}{\sqrt{5}} & 0 \end{bmatrix} \begin{bmatrix} 0 & -1 & 0 \\ -1 & 0 & 0 \\ 0 & 0 & 1 \end{bmatrix} \begin{bmatrix} \frac{2}{\sqrt{5}} \\ \frac{1}{\sqrt{5}} \\ 0 \end{bmatrix} \begin{bmatrix} \frac{1}{\sqrt{5}} \\ \frac{-2}{\sqrt{5}} \\ 0 \end{bmatrix}$$

$$= \begin{bmatrix} \frac{2}{\sqrt{5}} \\ \frac{1}{\sqrt{5}} \\ 0 \end{bmatrix} + \frac{3\epsilon}{100} \begin{bmatrix} \frac{1}{\sqrt{5}} \\ \frac{-2}{\sqrt{5}} \\ 0 \end{bmatrix} = \begin{bmatrix} 2 + \frac{3\epsilon}{20} \\ 1 \\ 0 \end{bmatrix}$$

$$\tilde{x}^2 \approx \begin{bmatrix} 0 \\ 0 \\ 1 \end{bmatrix} + \frac{\epsilon}{20}\begin{bmatrix} \frac{1}{\sqrt{5}} & \frac{-2}{\sqrt{5}} & 0 \end{bmatrix}\begin{bmatrix} 0 & -1 & 0 \\ -1 & 0 & 0 \\ 0 & 0 & 1 \end{bmatrix}\begin{bmatrix} 0 \\ 0 \\ 1 \end{bmatrix}\begin{bmatrix} \frac{1}{\sqrt{5}} \\ \frac{-2}{\sqrt{5}} \\ 0 \end{bmatrix}$$

$$= \begin{bmatrix} 0 \\ 0 \\ 1 \end{bmatrix}$$

which are the same results as obtained before.

Now we come to the last case where λ is a non-semi-simple eigenvalue of multiplicity m; we proceed to calculate the shifts in λ due to small perturbations in A equal to ϵA_1. Recalling the Puiseux expansion, the eigenvalue $\tilde{\lambda}$ and corresponding eigenvectors $\tilde{x}$ of $A + \epsilon A_1$ are represented in the form

$$\tilde{\lambda}_i = \lambda + \lambda_i^{(1)}\sqrt[m]{\epsilon} + \lambda_i^{(2)}\sqrt[m]{\epsilon^2} + \cdots + \lambda_i^{(m)}\epsilon + \lambda_i^{(m+1)}\sqrt[m]{\epsilon^{m+1}} + \cdots, \quad i = 1, \ldots, m$$

and

$$\tilde{x}^i = x^i + x^i_{(1)}\sqrt[m]{\epsilon} + x^i_{(2)}\sqrt[m]{\epsilon^2} + \cdots + x^i_{(m)}\epsilon + x^i_{(m+1)}\sqrt[m]{\epsilon^{m+1}} + \cdots, i = 1, \ldots, m$$

By substituting in the eigenvalue problem for the perturbed matrix

$$(A + \epsilon A_1)(x^i + x^i_{(1)}\sqrt[m]{\epsilon} + \cdots) = (\lambda + \lambda_i^{(1)}\sqrt[m]{\epsilon} + \cdots)(x^i + x^i_{(1)}\sqrt[m]{\epsilon} + \cdots)$$

and grouping the same powers of ϵ, we obtain

$$Ax^i_{(1)} = \lambda x^i_{(1)} + \lambda_i^{(1)}x^i$$

$$Ax^i_{(2)} = \lambda x^i_{(2)} + \lambda_i^{(1)}x^i_{(1)} + \lambda_i^{(2)}x^i$$

$$\vdots$$

$$Ax^i_{(m)} + A_1x^i = \lambda x^i_{(m)} + \lambda_i^{(1)}x^i_{(m-1)} + \cdots + \lambda_i^{(m)}x^i$$

$$\vdots$$

So in order to calculate $x^i_{(1)}$ we need to calculate first $\lambda_i^{(1)}$; to calculate $x^i_{(2)}$ we need to know $x^i_{(1)}$, $\lambda_i^{(1)}$ and $\lambda_i^{(2)}$, and so on. Therefore we give a method to determine $\lambda_i^{(1)}$ and $u^i_{(1)}$ and the interested reader can follow the same method to calculate the rest.

$\lambda_i^{(1)}$ and $\lambda_i^{(m)}$ have been calculated before and we give them again:

$$\lambda_i^{(1)} = \sqrt[m]{-\frac{m!D_1(\lambda)}{D_0^{(m)}(\lambda)}}\, e^{2ji\pi/m}, \quad j = \sqrt{-1}, \quad i = 1, \ldots, m$$

and

$$\lambda_i^{(m)} = -\frac{1}{m}\left(\sum_{j=m+1}^{n} \lambda_j^{(1)} + \frac{1}{(n-1)!}D_1^{(n-1)}(\lambda)\right), \quad i = 1, \ldots, m$$

However we wish to give direct formulae for calculating $\lambda_i^{(1)}$ and $\lambda_i^{(m)}$ better than from calculating the polynomial $|\lambda I - A - \epsilon A_1|$. For this we use the expansion of the latter which was given before:

$$|\mu I - A - \epsilon A_1| = |\mu I - A|(1 - \epsilon\,\mathrm{tr}(\mu I - A)^{-1}A_1 + \cdots)$$

From this we obtain

$$D_1(\mu) = -\mathrm{tr}(\mu I - A)^a A_1 = -\mathrm{tr}\, T^{-1}(\mu I - A)^a T T^{-1} A_1 T$$

$$= -\mathrm{tr}\left[\begin{array}{cccc:ccc} \frac{|\mu I - A|}{\mu-\lambda} & \frac{|\mu I - A|}{(\mu-\lambda)^2} & \cdots & \frac{|\mu I - A|}{(\mu-\lambda)^m} & & & \\ & \frac{|\mu I - A|}{\mu-\lambda} & \ddots & \vdots & & 0 & \\ & 0 & & \frac{|\mu I - A|}{\mu-\lambda} & & & \\ \hdashline & & & & \frac{|\mu I - A|}{\mu-\lambda_{m+1}} & & \\ & 0 & & & & \ddots & \frac{|\mu I - A|}{\mu-\lambda_n} \end{array}\right] T^{-1}A_1 T$$

Therefore to calculate $D_1(\lambda)$ we substitute in the matrix above $\mu = \lambda$. The reader can check as before that

$$\left.\frac{|\mu I - A|}{\mu - \lambda_i}\right|_{\mu=\lambda} = 0, \quad i = m+1, \ldots, n$$

Also that

$$\left.\frac{|\mu I - A|}{\mu - \lambda}\right|_{\mu=\lambda} = \left.\frac{|\mu I - A|}{(\mu - \lambda)^2}\right|_{\mu=\lambda} = \cdots = \left.\frac{|\mu I - A|}{(\mu - \lambda)^{m-1}}\right|_{\mu=\lambda} = 0$$

and that

$$\left.\frac{|\mu I - A|}{(\mu - \lambda)^m}\right|_{\mu=\lambda} = \prod_{i=m+1}^{n}(\lambda - \lambda_i) = \frac{D_0^{(m)}(\lambda)}{m!}$$

Therefore we obtain

$$D_1(\lambda) = -\frac{D_0^{(m)}(\lambda)}{m!}\langle v^m, A_1 u^1\rangle$$

and by substituting in $\lambda_i^{(1)}$ we have

$$\lambda_i^{(1)} = \sqrt[m]{\frac{-m!D_1(\lambda)}{D_0^{(m)}(\lambda)}}\, e^{2ji\pi/m} = \sqrt[m]{\langle v^m, A_1 u^1\rangle}\, e^{2ji\pi/m}, \quad j = \sqrt{-1}, \quad i = 1, \ldots, m$$

To compute $\lambda_i^{(m)}$ we first obtain $D_1^{(n-1)}(\lambda)$. It can be easily checked that

$$\frac{d^{n-1}}{d\mu^{n-1}} \frac{|\mu I - A|}{\mu - \lambda} = (n-1)!$$

and the rest are all zero. Therefore

$$D_1^{(n-1)}(\lambda) = -(n-1)! \operatorname{tr} A_1$$

and we obtain

$$\lambda_i^{(m)} = -\frac{1}{m}\left(\sum_{j=m+1}^{n} \lambda_j^{(1)} - \operatorname{tr} A_1 \right)$$

Substituting for $\lambda_j^{(1)}$ from theorem 1, we arrive at the final result:

$$\lambda_i^{(m)} = -\frac{1}{m}\left(\sum_{j=m+1}^{n} \langle v^j, A_1 u^j \rangle - \operatorname{tr} A_1 \right)$$

$$= -\frac{1}{m}\left(\sum_{j=m+1}^{n} \langle v^j, A_1 u^j \rangle - \operatorname{tr} T^{-1} A_1 T \right)$$

$$= -\frac{1}{m}\left(\sum_{j=m+1}^{n} \langle v^j, A_1 u^j \rangle - \sum_{j=1}^{n} \langle v^j, A_1 u^j \rangle \right)$$

$$= \frac{1}{m} \sum_{j=1}^{m} \langle v^j, A_1 u^j \rangle$$

In this fascinating book, Kato (1976), p. 79, obtained the same formula for $\lambda_i^{(m)}$, which he called the *weighted mean* of the eigenvalues, through integrating the resolvent around a closed contour containing the eigenvalues.

To compute $x_{(1)}^i$ we solve the eigenvalue problem given before:

$$Ax_{(1)}^i = \lambda x_{(1)}^i + \lambda_i^{(1)} x^i$$

i.e.

$$(A - \lambda I)x_{(1)}^i = \lambda_i^{(1)} x^i$$

From this we obtain according to Section 4.7, and assuming that A is non-derogatory,

$$x_{(1)}^i = \lambda_i^{(1)} u^2$$

Therefore we have established

THEOREM 5. If λ is a non-semi-simple eigenvalue of multiplicity m of a matrix A, with corresponding generalized eigenvectors $u^1, u^2, \ldots, u^m$, the eigenvalues $\tilde{\lambda}_i, i = 1, \ldots, m$ of $A + \epsilon A_1$ will lie on the circumference of a circle with centre z and radius $r \approx |\sqrt[m]{\epsilon}\, \lambda_i^{(1)}|$, where

$$z \approx \lambda + \frac{1}{m} \sum_{j=1}^{m} \langle v^j, \epsilon A_1 u^j \rangle$$

and

$$\lambda_i^{(1)} = \sqrt[m]{\langle v^m, A_1 u^1 \rangle}\, e^{2ji\pi/m}, \quad j = \sqrt{-1}, \quad i = 1, \ldots, m$$

where $v^1, \ldots, v^m$ are the reciprocal bases for $u^1, \ldots, u^m$. And the eigenvector $\tilde{x}^i$ of $A + \epsilon A_1$ is given by

$$\tilde{x}^i \approx x^i + \sqrt[m]{\epsilon}\, \lambda_i^{(1)} u^2$$

where x^i is the eigenvector of A, and we assume that A is non-derogatory without loss of generality.

Example. Let A and A_1 be

$$A = \begin{bmatrix} 17 & 0 & -25 \\ 0 & 3 & 0 \\ 9 & 0 & -13 \end{bmatrix} \quad A_1 = \begin{bmatrix} 1 & 0 & -1 \\ 0 & 1 & 0 \\ 0 & 0 & 2 \end{bmatrix}$$

The eigenvalues of A are 3, 2, 2.
The generalized eigenvectors of A are

$$\begin{bmatrix} 0 \\ 1 \\ 0 \end{bmatrix}, \quad \begin{bmatrix} 5 \\ 0 \\ 3 \end{bmatrix}, \quad \begin{bmatrix} 1/3 \\ 0 \\ 0 \end{bmatrix}$$

Their reciprocal bases are

$$\begin{bmatrix} 0 \\ 1 \\ 0 \end{bmatrix}, \quad \begin{bmatrix} 0 \\ 0 \\ 1/3 \end{bmatrix}, \quad \begin{bmatrix} 3 \\ 0 \\ -5 \end{bmatrix}$$

The eigenvalues of $A + \epsilon A_1$ are the roots of

$$\begin{vmatrix} 17 + \epsilon - \lambda & 0 & -25 - \epsilon \\ 0 & 3 + \epsilon - \lambda & 0 \\ 9 & 0 & -13 + 2\epsilon - \lambda \end{vmatrix} = 0$$

giving the new eigenvalues of the perturbed matrix

$$\tilde{\lambda}_1 = 3 + \epsilon$$

$$\tilde{\lambda}_2 = 2 + \frac{3}{2}\epsilon + i\sqrt{24\epsilon - \tfrac{1}{4}\epsilon^2}$$

$$\tilde{\lambda}_3 = 2 + \frac{3}{2}\epsilon - i\sqrt{24\epsilon - \tfrac{1}{4}\epsilon^2}$$

The new eigenvectors are obtained by solving the equations $(\tilde{\lambda} I - A - \epsilon A_1)\tilde{u} = 0$

to give

$$\tilde{u}^2 = \begin{bmatrix} 5 - \frac{1}{6}\epsilon + i\frac{\sqrt{24\epsilon}}{3} \\ 0 \\ 3 \end{bmatrix}, \qquad \tilde{u}^3 = \begin{bmatrix} 5 - \frac{1}{6}\epsilon - i\frac{\sqrt{24\epsilon}}{3} \\ 0 \\ 3 \end{bmatrix}$$

And by using the perturbation technique we obtain

$$\tilde{\lambda} \approx 2 + \frac{\epsilon}{2}\left[\langle v^2, A_1 u^2\rangle + \langle v^3, A_1 u^3\rangle\right] + \sqrt{\langle v^3, \epsilon A_1 u^2\rangle}$$

where

$$\langle v^2, A_1 u^2\rangle + \langle v^3, A_1 u^3\rangle = [0 \quad 0 \quad 1/3]\begin{bmatrix} 1 & 0 & -1 \\ 0 & 1 & 0 \\ 0 & 0 & 2 \end{bmatrix}\begin{bmatrix} 5 \\ 0 \\ 3 \end{bmatrix}$$

$$+ [3 \quad 0 \quad -5]\begin{bmatrix} 1 & 0 & -1 \\ 0 & 1 & 0 \\ 0 & 0 & 2 \end{bmatrix}\begin{bmatrix} 1/3 \\ 0 \\ 0 \end{bmatrix} = 3$$

and

$$\langle v^3, A_1 u^2\rangle = [3 \quad 0 \quad -5]\begin{bmatrix} 1 & 0 & -1 \\ 0 & 1 & 0 \\ 0 & 0 & 2 \end{bmatrix}\begin{bmatrix} 5 \\ 0 \\ 3 \end{bmatrix} = -24$$

Therefore

$$\tilde{\lambda}_2 \approx 2 + \frac{3\epsilon}{2} + i\sqrt{24\epsilon}$$

$$\tilde{\lambda}_3 \approx 2 + \frac{3\epsilon}{2} - i\sqrt{24\epsilon}$$

and also

$$\tilde{x}^2 \approx x^2 + \lambda_2^{(1)}\sqrt{\epsilon}\, u^3 = \begin{bmatrix} 5 \\ 0 \\ 3 \end{bmatrix} + i\sqrt{24}\begin{bmatrix} 1/3 \\ 0 \\ 0 \end{bmatrix}\sqrt{\epsilon} = \begin{bmatrix} 5 + \frac{i\sqrt{24\epsilon}}{3} \\ 0 \\ 3 \end{bmatrix}$$

$$\tilde{x}^3 \approx x^3 + \lambda_3^{(1)}\sqrt{\epsilon}\, u^3 = \begin{bmatrix} 5 \\ 0 \\ 3 \end{bmatrix} - i\sqrt{24}\begin{bmatrix} 1/3 \\ 0 \\ 0 \end{bmatrix}\sqrt{\epsilon} = \begin{bmatrix} 5 - \frac{i\sqrt{24\epsilon}}{3} \\ 0 \\ 3 \end{bmatrix}$$

which are the same results obtained above for small enough $|\epsilon|$.

6.4. Perturbations in the exponential of a matrix

Unlike the eigenvalues and eigenvectors, the perturbations in e^A are the same whether A is semi-simple or not. In this section we calculate $e^{A+\epsilon A_1}$, very useful in studying linear differential equations under variations occurring in the state matrix. The reader knows from before that if $AA_1 = A_1A$ then

$$e^{A+\epsilon A_1} = e^A \cdot e^{\epsilon A_1}$$

$$= e^A \left(I + \epsilon A_1 + \frac{\epsilon^2 A_1^2}{2!} + \cdots \right)$$

but unfortunately $AA_1 \neq A_1A$ in general. To find $e^{A+\epsilon A_1}$, when A and A_1 are not commutative, we use an iterative technique as follows:

We know that $e^{A+\epsilon A_1}$ is the solution of

$$\frac{dX}{dt} = (A + \epsilon A_1)X, \quad X(0) = I$$

when substituting $t = 1$. However the above differential equation can also be written in the following form

$$\frac{dX}{dt} = AX + \epsilon A_1 X, \quad X(0) = I$$

whose solution is given from before as

$$X = e^{At} + e^{At} \int_0^t e^{-A\tau} \epsilon A_1 X \, d\tau,$$

By iteration we obtain

$$X = e^{At} + e^{At} \int_0^t e^{-A\tau} \epsilon A_1 \left(e^{A\tau} + e^{A\tau} \int_0^\tau e^{-As} \epsilon A_1 X \, ds \right) d\tau$$

$$= e^{At} + \epsilon e^{At} \int_0^t e^{-A\tau} A_1 e^{A\tau} \, d\tau + \epsilon^2 e^{At} \int_0^t e^{-A\tau} A_1 e^{A\tau} \int_0^\tau e^{-As} A_1 X \, ds \, d\tau$$

Substituting $t = 1$ and comparing results, we obtain

$$e^{A+\epsilon A_1} = e^A + \epsilon e^A \int_0^1 e^{-A\tau} A_1 e^{A\tau} \, d\tau + \cdots$$

Example. Compute $e^{A+\epsilon A_1}$ if

$$A = \begin{bmatrix} 2 & -1 \\ 1 & 0 \end{bmatrix}, \quad A_1 = \begin{bmatrix} 0 & 1 \\ 0 & 0 \end{bmatrix}$$

$$e^{A+\epsilon A_1} \approx e^A + \epsilon e^A \int_0^1 e^{-A\tau} A_1 e^{A\tau}\, d\tau$$

$$= \begin{bmatrix} 2e & -e \\ e & 0 \end{bmatrix} + \epsilon \begin{bmatrix} 2e & -e \\ e & 0 \end{bmatrix} \int_0^1 \begin{bmatrix} -\tau e^{-\tau} + e^{-\tau} & \tau e^{-\tau} \\ -\tau e^{-\tau} & \tau e^{-\tau} + e^{-\tau} \end{bmatrix} \begin{bmatrix} 0 & 1 \\ 0 & 0 \end{bmatrix} \begin{bmatrix} \tau e^{\tau} + e^{\tau} & -\tau e^{\tau} \\ \tau e^{\tau} & e^{\tau} - \tau e^{\tau} \end{bmatrix} d\tau$$

$$= \begin{bmatrix} 2e & -e \\ e & 0 \end{bmatrix} + \epsilon \begin{bmatrix} 2e & -e \\ e & 0 \end{bmatrix} \int_0^1 \begin{bmatrix} -\tau^2 + \tau & \tau^2 - 2\tau + 1 \\ -\tau^2 & -\tau + \tau^2 \end{bmatrix} d\tau$$

$$= \begin{bmatrix} 2e & -e \\ e & 0 \end{bmatrix} + \epsilon \begin{bmatrix} 2e & -e \\ e & 0 \end{bmatrix} \cdot \begin{bmatrix} 1/6 & 1/3 \\ -1/3 & -1/6 \end{bmatrix}$$

$$= \begin{bmatrix} 2e & -e \\ e & 0 \end{bmatrix} + \epsilon \begin{bmatrix} \frac{2}{3}e & \frac{5}{6}e \\ \frac{e}{6} & \frac{1}{3}e \end{bmatrix}$$

As an application of $e^{A+\epsilon A_1}$ in the field of differential equations, assume that in the linear first-order differential equation

$$\frac{dx}{dt} = Ax + u(t), \quad x(0) = x^0,$$

A and u have both exhibited first-order variations of the form $A + \epsilon A_1$ and $u(t) + \epsilon u^1(t)$, respectively, and we study their effect on $x(t)$. Writing

$$\frac{d\tilde{x}}{dt} = (A + \epsilon A_1)\tilde{x} + \tilde{u}(t)$$

i.e.

$$\frac{d}{dt}(x + \epsilon x^1) = (A + \epsilon A_1)(x + \epsilon x^1) + u + \epsilon u^1$$

and grouping terms in ϵ, we obtain

$$\frac{dx^1}{dt} = Ax^1 + A_1 x + u^1$$

whose solution, neglecting initial conditions of x^1, is directly obtained:

$$x^1(t) = e^{At} \int_0^t e^{-A\tau}(A_1 x(\tau) + u^1(\tau))\, d\tau$$

and where $x(t)$ is the solution of the unperturbed system. The reader should arrive at the same result by writing

$$\tilde{x}(t) = x(t) + \epsilon x^1(t) = e^{(A+\epsilon A_1)t}\tilde{x}(0) + e^{(A+\epsilon A_1)t} \int_0^t e^{-(A+\epsilon A_1)\tau}(u(\tau) + \\ + \epsilon u^1(\tau))\,\mathrm{d}\tau$$

6.5. Sensitivity analysis

Sensitivity analysis is one of the most useful tools in the design, control and measurement of many engineering systems. If the system is represented by a set of equations (linear, nonlinear, differential), the engineer is usually interested in knowing whether his system is sensitive or not to some changes in the system's parameters, and if the system is sensitive, how he is going to measure its sensitivity. Sensitivity analysis is a wide discipline in which some of the topics are trivial and some highly complicated. In this section we study only eigenvalue sensitivity of a matrix to changes in the matrix elements. The method can be applied similarly to eigenvector sensitivity or to solution sensitivity of differential and integral equations.

Consider a matrix A of order n, where each element a_{ij} is a function of many variables $x_1, x_2, \ldots, x_m$, and where it is required to calculate $\dfrac{\partial \lambda_i}{\partial x_k}$. The latter is calculated from the formula

$$\frac{\partial \lambda_i}{\partial x_k} = \lim_{\Delta x_k \to 0} \frac{\lambda_i(x_1, \ldots, x_k + \Delta_k, \ldots, x_m) - \lambda_i(x_1, \ldots, x_k, \ldots, x_m)}{\Delta x_k}$$

We will assume, for simplicity, that λ_i is a distinct eigenvalue of A, and the reader can obtain a similar result if λ_i is a multiple eigenvalue. The matrix A can be expanded by Taylor's series around Δx_k in the following manner

$$A(x_1, \ldots, x_k + \Delta x_k, \ldots, x_m) = A(x_1, \ldots, x_k, \ldots, x_m) + \\ + \Delta x_k A_1(x_1, \ldots, x_k, \ldots, x_m) + (\Delta x_k)^2 A_2(x_1, \ldots, x_k, \ldots, x_m) + \cdots$$

where A_1 is obtained from A by differentiating all its elements with respect to x_k. Therefore applying the concept of perturbation on the eigenvalue of the perturbed matrix

$$A(x_1, \ldots, x_k + \Delta x_k, \ldots, x_m),$$

we obtain

$$\lambda_i(x_1, \ldots, x_k + \Delta x_k, \ldots, x_m) = \lambda_i(x_1, \ldots, x_k, \ldots, x_m) + \\ + \Delta x_k \langle v^i, A_1 u^i \rangle + 0(\Delta x_k)^2$$

and by substituting in

$$\frac{\partial \lambda_i}{\partial x_k}$$

we obtain

$$\frac{\partial \lambda_i}{\partial x_k} = \langle v^i, A_1 u^i \rangle$$

where u^i is the eigenvector of A corresponding to λ_i and v^i is its reciprocal vector.

Example. Let

$$A = \begin{bmatrix} 6 & (2+x)^3 \\ 8e^{y+x} & -6 \end{bmatrix}$$

and it is required to calculate

$$\left.\frac{\partial \lambda}{\partial x}\right|_{x=0, y=0}$$

The matrix A_1 is given by

$$\left.\frac{\partial A}{\partial x}\right|_{x=0, y=0} = \begin{bmatrix} 0 & 12 \\ 8 & 0 \end{bmatrix}$$

and

$$A|_{x=0, y=0} = \begin{bmatrix} 6 & 8 \\ 8 & -6 \end{bmatrix}$$

whose eigenvalues are 10, −10 and whose eigenvectors are

$$\begin{bmatrix} \frac{2}{\sqrt{5}} \\ \frac{1}{\sqrt{5}} \end{bmatrix}, \quad \begin{bmatrix} \frac{1}{\sqrt{5}} \\ \frac{-2}{\sqrt{5}} \end{bmatrix}$$

which are the same as their reciprocals. Therefore

$$\left.\frac{\partial \lambda_1}{\partial x}\right|_{x=y=0} = \begin{bmatrix} \frac{2}{\sqrt{5}} & \frac{1}{\sqrt{5}} \end{bmatrix} \begin{bmatrix} 0 & 12 \\ 8 & 0 \end{bmatrix} \begin{bmatrix} \frac{2}{\sqrt{5}} \\ \frac{1}{\sqrt{5}} \end{bmatrix} = \frac{16}{5} + \frac{24}{5} = \frac{40}{5} = 8$$

For application of sensitivity analysis to network design, the reader is referred to Papoulis (1966), Sobhy and Deif (1975, 1976), and Deif (1981).

Exercises 6.5

1. Show that the shift occurring in a distinct eigenvalue λ_i of a matrix A due to variation ϵA_1 in A is equal to

$$\epsilon \operatorname{tr}\left(A_1 \prod_{k \neq i} \frac{A - \lambda_k I}{\lambda_i - \lambda_k} \right)$$

Hint: use Exercise 4.10.48.

2. If λ_i is a distinct eigenvalue of a non-semi-simple matrix A, with corresponding eigenvector u^i, obtain the eigenvalue $\tilde{\lambda}_i$ and eigenvector $\tilde{u}$ of $A + \epsilon A_1$ to a first-order approximation. *Hint*: use a proof similar to that of Theorem 1, Section 6.3, while using the generalized eigenvectors of A.
3. If λ is a semi-simple eigenvalue of multiplicity m of a non-semi-simple matrix A, with eigenvectors $u^1, u^2, \ldots, u^m$, obtain to a first-order approximation the eigenvalues $\tilde{\lambda}_1, \tilde{\lambda}_2, \ldots, \tilde{\lambda}_m$ and corresponding eigenvectors $\tilde{u}^1, \ldots, \tilde{u}^m$ of $A + \epsilon A_1$.
4. Show that the eigenvalue $\tilde{\lambda}_i$ of $A + \epsilon A_1$ with A semi-simple lies inside a circle of centre λ_i of A and radius equal to $\mathscr{K}(A)\| \epsilon A_1 \|$, where $\mathscr{K}(A)$ is the spectral condition number of A.
5. If A is Hermitian show that $\tilde{\lambda}_i(A + \epsilon A_1) \approx \lambda_i + \langle u^i, \epsilon A_1 u^i \rangle$, if λ_i is a distinct eigenvalue of A, and u^i is the normalized eigenvector of A corresponding to λ_i.
6. If A and A_1 commutes, show that $\tilde{\lambda}_i(A + \epsilon A_1) = \lambda_i(A) + \epsilon\lambda_i(A_1)$.
7. Show that the sum of the deviations of order ϵ in the eigenvalues of A due to perturbation in A equal to ϵA_1 are equal to ϵ tr A_1.
8. Obtain the eigenvalues and eigenvectors of $A + \epsilon A_1$ using perturbation techniques if

$$A = \begin{bmatrix} -4 & -4 \\ 1 & 0 \end{bmatrix} \quad \text{and } A_1 = \begin{bmatrix} 1 & -1 \\ 0 & 2 \end{bmatrix}.$$

9. Consider the eigenvalue problem $(M\lambda_i^2 + B\lambda_i + K)u^i = 0$, where M, B and K are all positive semidefinite. Show that if λ_i is a distinct root of $| M\lambda^2 + B\lambda + K | = 0$ the root $\tilde{\lambda}_i$ of $| (M + \epsilon M_1)\tilde{\lambda}^2 + (B + \epsilon B_1)\tilde{\lambda} + K + \epsilon K_1 | = 0$ is given by $\tilde{\lambda}_i \approx \lambda_i + \epsilon\lambda_i^{(1)}$, where

$$\lambda_i^{(1)} \approx -\frac{\langle u^i, (M_1\lambda_i^2 + B_1\lambda_i + K_1)u^i \rangle}{\langle u^i, (2M\lambda_i + B)u^i \rangle}$$

Show also that the eigenvector $\tilde{u}^i$ is given by $\tilde{u}^i \approx u^i + \epsilon u^i_{(1)}$, where

$$u^i_{(1)} = \sum_{k=1}^{n} c_{ik} u^k$$

where n is the dimension of the matrices, and c_{ik} are obtained from solving the set of linear equations:

$$\langle u^k, (2M\lambda_i + B)u^i \rangle \lambda_i^{(1)} + \langle u^k, (M_1\lambda_i^2 + B_1\lambda_i + K_1)u^i \rangle +$$
$$+ \langle u^k, (M\lambda_i^2 + B\lambda_i + K) \sum_{\substack{k=1 \\ k \neq i}}^{n} c_{ik}u^k \rangle = 0, \quad k = 1, \ldots, n$$

10. If in the differential equation

$$M\frac{d^2x}{dt^2} + B\frac{dx}{dt} + Kx = F(t)$$

M, B and K exhibit first-order variations of the form ϵM_1, ϵB_1 and ϵK_1, then show that the solution of the differential equation exhibits first-order variation given by:

$$\epsilon x^t_{(1)} = -\epsilon\,(MD^2 + BD + K)^a \sum_{i=1}^{n} \frac{e^{\lambda_i t}}{\Delta(\lambda_i)} \int_0^t e^{-\lambda_i \tau}(M_1\ddot{x}(\tau) + B_1\dot{x}(\tau) + K_1 x(\tau))\, d\tau$$

where $x(t)$ is the solution of the unperturbed system, λ_i is a root of

$$|M\lambda^2 + B\lambda + K| = 0, \quad D = \frac{\mathrm{d}}{\mathrm{d}t}$$

and

$$\Delta(\lambda_i) = \lim_{\lambda \to \lambda_i} \frac{|M\lambda^2 + B\lambda + K|}{\lambda - \lambda_i}$$

11. Show that $\operatorname{tr} e^{A+\epsilon A_1} = \operatorname{tr} e^A + \epsilon \operatorname{tr}(e^A A_1) + 0(\epsilon^2)$.
12. Obtain a solution of the differential equation

$$\frac{\mathrm{d}x}{\mathrm{d}t} = (A(t) + \epsilon B(t))x + u(t)$$

in powers of ϵ.
13. Obtain in powers of ϵ, $|e^{A+\epsilon A_1}|$.
14. Show that

$$\frac{\mathrm{d}\lambda(A^{-1})}{\mathrm{d}\alpha} = -\frac{1}{\lambda^2(A)} \cdot \frac{\mathrm{d}\lambda(A)}{\mathrm{d}\alpha}$$

15. If in the integral equation

$$x(t) = f(t) + \lambda \int_a^b k(\epsilon, t) x(\epsilon)\, \mathrm{d}\epsilon$$

the kernel $k(\epsilon, t)$ exhibits a perturbation of the form $\epsilon k_1(\epsilon, t)$, obtain the solution of the perturbed integral equation in the form $\tilde{x}(t) = x(t) + \epsilon x_1(t) + \epsilon^2 x_2(t) + \cdots$.
16. Consider the eigenvalue problem $Ae^{\tau s}u = Bu$, with A nonsingular. Show that if the matrices A and B have exhibited variations of the form $A + \epsilon A_1$, $B + \epsilon B_1$, and the eigenvalues of the unperturbed system are distinct, then the shifts in the eigenvalues and eigenvectors due to the above perturbations are given by $\tilde{s}_i \approx s_i + \epsilon s_i^{(1)}$, $\tilde{u}^i \approx u^i + \epsilon u^i_{(1)}$, $i = 1, \ldots, n$, where

$$s_i^{(1)} = -\frac{\langle v^i, (A_1 e^{\tau s_i} - B_1)u^i \rangle}{\tau e^{\tau s_i}}, \quad u^i_{(1)} = -\sum_{\substack{k=1 \\ k \neq i}}^{n} \frac{\langle v^k, (A_1 e^{\tau s_i} - B_1)u^i \rangle}{(e^{\tau s_i} - e^{\tau s_k})} u^k$$

and where $\langle v^i, Au^j \rangle = \delta_{ij}$. For a proof and applications see Deif and Selim (1981).
17. Consider the eigenvalue problem $(A + \lambda B)u = 0$, where A and B are both Hermitian and B is positive definite. Show that, if A and B exhibit variations of the form $A + \epsilon A_1$ and $B + \epsilon B_1$, with A_1 and B_1 Hermitian, the shift in the eigenvalue λ_i is given by

$$\epsilon\lambda_i^{(1)} = -\epsilon \frac{\langle u^i, (A_1 + \lambda_i B_1)u^i \rangle}{\langle u^i, Bu^i \rangle}$$

where λ_i is a distinct eigenvalue and u^i its corresponding eigenvector. Obtain also the shift in the eigenvector u^i. Show also, in the case of degeneracy where λ is an eigenvalue of multiplicity m, with corresponding eigenvectors $z^1, z^2, \ldots, z^m$, that

$\lambda_i^{(1)}$ $(i = 1, \ldots, m)$ are the roots of

$$\begin{vmatrix} \langle z^1, (\lambda^{(1)}B + A_1 + \lambda B_1) z^1 \rangle & \cdots & \langle z^m, (\lambda^{(1)}B + A_1 + \lambda B_1) z^1 \rangle \\ \hdashline \langle z^1, (\lambda^{(1)}B + A_1 + \lambda B_1) z^m \rangle & \cdots & \langle z^m, (\lambda^{(1)}B + A_1 + \lambda B_1) z^m \rangle \end{vmatrix} = 0$$

Obtain the shifts in the eigenvectors z^i $(i = 1, \ldots, m)$. Discuss the case when B is positive semidefinite.

18. Show that if A is semi-simple, then $e^{A+\epsilon B}$ can be obtained easily from the formula

$$e^{A+\epsilon B} \approx e^A + \epsilon TWT^{-1}$$

where T is the modal matrix of A,

$$w_{ij} = \frac{e^{\lambda_j} - e^{\lambda_i}}{\lambda_j - \lambda_i} \langle v^i, Bu^j \rangle$$

and where λ_i is an eigenvalue of A, u^j is an eigenvector ot A and v^i is a reciprocal eigenvector of A. Obtain a formula for A non-semi-simple.

19. If A and B are functions of scalars $\alpha_1, \ldots, \alpha_m$, and

$$\frac{dx}{dt} = Ax + Bu(t)$$

show that

$$\frac{\partial x}{\partial \alpha_k} = e^{At} \int_0^t e^{-A\tau} \left(\frac{\partial A}{\partial \alpha_k} x(\tau) + \frac{\partial B}{\partial \alpha_k} u(\tau) \right) d\tau.$$

20. In certain electrical applications, one may require the sensitivity of the response on the jw axis due to variations in the network variables $\alpha_1, \ldots, \alpha_m$. If the differential equation of the network is formulated as

$$\frac{dx}{dt} = Ax + Bu(t)$$

where A and B contain $\alpha_1, \ldots, \alpha_m$, and the response of the unperturbed system is given by

$$\left| \frac{c^T x(s)}{u(s)} \right| = |R(w)|$$

where c^T is a row vector relating the output to the state variables, show that

$$\frac{\partial |R(w)|}{\partial \alpha_k} = \frac{R^r \tilde{R}_k^r + R^i \tilde{R}_k^i}{|R(w)|}$$

where r and i refer to the real and imaginary parts of R, and

$$\tilde{R} = \left. \frac{c^T \tilde{x}_k(s)}{u(s)} \right|_{s=jw}$$

where

$$\tilde{x}_k(s) = (sI - A)^{-1} \left(\frac{\partial A}{\partial \alpha_k} x(s) + \frac{\partial B}{\partial \alpha_k} u(s) \right).$$

APPENDIX ONE

THE HOLDER INEQUALITY

The Holder inequality

$$\sum_{i=1}^{n} |a_i||b_i| \leqslant \left(\sum_{i=1}^{n} |a_i|^p\right)^{1/p} \left(\sum_{i=1}^{n} |b_i|^q\right)^{1/q}$$

where

$$\frac{1}{p} + \frac{1}{q} = 1, \quad \text{and } p > 1$$

Proof. Consider the curve $v = u^{p-1}$.

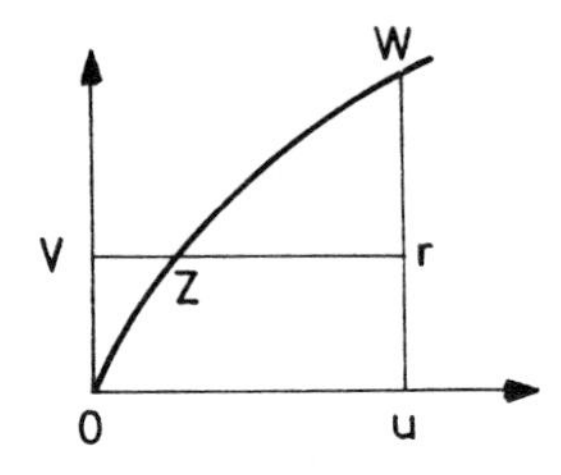

It is obvious that the area $0vru \leqslant$ area $0zv$ + area $0wu$, where the letter symbols u and v represent points on the u and v axes. They will become equal when z coincides with w.
Therefore

$$uv \leqslant \int_0^v v^{1/p-1}\, dv + \int_0^u u^{p-1}\, du$$

$$= \frac{v^q}{q} + \frac{u^p}{p}$$

Now let

$$u = |a_i| \Big/ \left(\sum_{i=1}^{n} |a_i|^p\right)^{1/p}$$

and

$$v = |b_i| \Big/ \left(\sum_{i=1}^{n} |b_i|^q \right)^{1/q}$$

so we obtain

$$\frac{|a_i|}{\left(\sum_{i=1}^{n} |a_i|^p \right)^{1/p}} \cdot \frac{|b_i|}{\left(\sum_{i=1}^{n} |b_i|^q \right)^{1/q}} \leqslant \frac{|a_i|^p}{p \sum_{i=1}^{n} |a_i|^p} + \frac{|b_i|^q}{q \sum_{i=1}^{n} |b_i|^q}$$

and by summing over i from 1 to n we obtain

$$\frac{\sum_{i=1}^{n} |a_i||b_i|}{\left(\sum_{i=1}^{n} |a_i|^p \right)^{1/p} \left(\sum_{i=1}^{n} |b_i|^q \right)^{1/q}} \leqslant \frac{\sum_{i=1}^{n} |a_i|^p}{p \sum_{i=1}^{n} |a_i|^p} + \frac{\sum_{i=1}^{n} |b_i|^q}{q \sum_{i=1}^{n} |b_i|^q}$$

and as

$$\frac{1}{p} + \frac{1}{q} = 1$$

we obtain directly the Holder inequality. Note that when $p = q = 2$ we obtain the special result called the Schwarz inequality.

$$\sum_{i=1}^{n} |a_i||b_i| \leqslant \sqrt{\sum_{i=1}^{n} |a_i|^2} \sqrt{\sum_{i=1}^{n} |b_i|^2}$$

The Holder inequality applies not only to variables but also to integrals; the reader can prove the following result

$$\int_R f(x)g(x)\,\mathrm{d}x \leqslant \left(\int_R f(x)^p\,\mathrm{d}x \right)^{1/p} \left(\int_R g(x)^q\,\mathrm{d}x \right)^{1/q}$$

where $f(x), g(x) \geqslant 0$. This can be done by letting

$$u = \frac{f(x)}{\left(\int_R f(x)^p\,\mathrm{d}x \right)^{1/p}}, \quad v = \frac{g(x)}{\left(\int_R g(x)^q\,\mathrm{d}x \right)^{1/q}}$$

and then by integrating the inequality in u and v w.r.t. x.

APPENDIX TWO

A PROPERTY OF POLYNOMIALS

A property of polynomials

For an nth order polynomial $\lambda^n + a_{n-1}\lambda^{n-1} + \cdots + a_0$, if

$$\sigma_r = \sum_{i=1}^{n} \lambda_i^r$$

where $\lambda_1, \lambda_2, \ldots, \lambda_n$ are the polynomial's roots, then we have

$$\sigma_k + \sigma_{k-1}a_{n-1} + \sigma_{k-2}a_{n-2} + \cdots + \sigma_1 a_{n-k+1} + k a_{n-k} = 0 \quad k = 1, 2, \ldots, n$$

Proof. Writing

$$\lambda^n + a_{n-1}\lambda^{n-1} + \cdots + a_0 = \prod_{i=1}^{n} (\lambda - \lambda_i)$$

we obtain upon equating coefficients of λ^{n-1}

$$a_{n-1} = -\sum_{i=1}^{n} \lambda_i = -\sigma_1$$

i.e.

$$\sigma_1 + a_{n-1} = 0$$

And upon squaring both sides and then equating coefficients of λ^{2n-2} we obtain

$$2a_{n-2} + a_{n-1}^2 = \sum_{i=1}^{n} \lambda_i^2 + 4\sum_{i,j} \lambda_i\lambda_j = \sigma_2 + 4a_{n-2}$$

Therefore

$$\sigma_2 + a_{n-2} - a_{n-1}^2 = 0$$

i.e.

$$\sigma_2 + a_{n-1}\sigma_1 + 2a_{n-2} = 0$$

And upon raising to the power of 3 for both sides and then equating coefficients of

λ^{3n-3} we obtain

$$3a_{n-3} + \sigma a_{n-1}a_{n-2} + a_{n-1}^3 = -\sum_{i=1}^{n} \lambda_i^3 - 9\sum_{i,j} \lambda_i\lambda_j \sum_{i=1}^{n} \lambda_i = \sigma_3 + 9a_{n-2}a_{n-1}$$

Therefore

$$\sigma_3 - 3a_{n-1}a_{n-2} + a_{n-1}^3 + 3a_{n-3} = 0$$

i.e.

$$\sigma_3 + a_{n-1}^3 - 2a_{n-1}a_{n-2} - a_{n-1}a_{n-2} + 3a_{n-3} = 0$$

or

$$\sigma_3 + a_{n-1}(a_{n-1}^2 - 2a_{n-2}) - a_{n-1}a_{n-2} + 3a_{n-3} = 0$$

i.e.

$$\sigma_3 + \sigma_2 a_{n-1} + \sigma_1 a_{n-2} + 3a_{n-3} = 0$$

Carrying the process further by rising both sides to the power 4, 5, . . . and then equating coefficients of λ^{4n-4}, λ^{5n-5}, etc., until λ^{kn-k} we obtain

$$\sigma_k + \sigma_{k-1}a_{n-1} + \cdots + a_{n-k+1}\sigma_1 + ka_{n-k} = 0$$

REFERENCES

Ayres, F.: *Theory and problems of matrices*. McGraw-Hill, Schaum's outline series, N.Y., 1975.

Barnes, J. G. P.: An algorithm for solving nonlinear equations based on the secant method, *Comput. J.*, **8**, p. 66, 1965.

Barnett, S. and Storey, C.: *Matrix methods in stability theory*. Nelson, London, 1970.

Bartels, R. H. and Stewart, G. W.: A solution of the equation AX + XB = C, *Commun. ACM*, **15**, pp. 820–826, 1972.

Bellman, R.: *Introduction to matrix analysis*. McGraw-Hill, N.Y., 1970.

Birkhoff, G. and Maclane, S.: *A survey of modern algebra*, Macmillan, N.Y., 1953.

Broyden, C. G.: A class of methods for solving non-linear equations, *Math. Comput.* **19**, p. 577, 1965.

Bunch, J. R. and Parlett, B. N.: Direct methods for solving symmetric indefinite systems of linear equations, *SIAM J. Numer Anal.*, **8**, pp. 639–655, 1971.

Burkill, J. C.: *The theory of ordinary differential equations*, Oliver and Boyd, Edinburgh, 1962.

Carnahan, B., Luther, H. and Wilkes, J.: *Applied numerical analysis*, John Wiley, N.Y., 1969.

Chan, S. P., Chan, S. Y. and Chan, S. G.: *Analysis of linear networks and systems*, Addison-Wesley, Reading, Mass., 1972.

Conte, S. D. and De Boor, C.: *Elementary numerical analysis; an algorithmic approach*, Third edit., McGraw-Hill, Tokyo, 1980.

Curtis, A. R. and Reid, J. K.: On the automatic scaling of matrices for Gaussian elimination, *J. Inst. Math. and its Appl.*, **10**, pp. 118–124, 1972.

Davidon, W. C.: *Variable metric method for minimization*, Argonne Nat. Lab., ANL – 5990 Rev., 1959.

Deif, A. S.: The proof of a new matrix theorem, *Proceedings of the Egyptian Math. and Phys. Soc.*, **51**, 1980.

Deif, A. S.: Optimal planning in education to meet social demand and manpower requirements, *Proc. Intern. Conf. Autom. Control.* IFAC, Cairo, 1977.

Deif, A. S.: A linear programming model for an optimum admissions policy in education, *Journ. Appl. Math. Modelling*, **4**, No. 1, 1980.

Deif, A. S.: Sensitivity analysis from the state equations by perturbation techniques, *Proc. AMS*, Lyon, 1981.

Deif, A. S. and Selim, F. I.: State space approach to first order perturbations in distributed systems, *Proc. ECCTD*, **81**, The Hague, 1981.

Delves, L. M. and Walsh, J.: *Numerical solution of integral equations*. Clarendon Press, Oxford, 1974.

Demidovich, B. P. and Maron, I. A.: *Computational mathematics*, Mir Publ., Moscow, 1973.

Derrick, W. R. and Grossman, S. I.: *Elementary differential equations with applications*. Addison-Wesley, Reading, Mass., 1976.

Desoer, C. A.: Perturbations of eigenvalues and eigenvectors of a network, *Proc. Fifth Allerton Conference on Circuit and System Theory*, pp. 8–11, 1967.
Faddeeva, V.: *Computational methods of linear algebra*. Dover, N.Y., 1959.
Finkbeiner, D. T.: *Introduction to matrices and linear transformations*, W. H. Freeman, San Francisco, 1960.
Fletcher, R.: *A review of methods for unconstrained optimization*, in an optimization symposium, University of Keele, England. Academic Press, London, 1969.
Fletcher, R.: Function minimization without evaluating derivatives; a review, *Comput. J.*, **8**, p. 33, 1965.
Fletcher, R.: Generalized inverse methods for the best least-squares solution of systems of nonlinear equations, *Comput. J.*, **10**, p. 392, 1968.
Fletcher, R. and Powell, M. J. D.: A rapidly convergent descent method for minimization, *Comput. J.*, **6**, p. 163, 1963.
Fletcher, R.: Methods for the solution of optimization problems, presented at the Symposium on Computer-aided Design Engineering, University of Waterloo, Canada, May, 1971.
Fletcher, R. and Reeves, C. M.: Function minimization by conjugate gradients, *Comput. J.*, **7**, p. 149, 1964.
Forsythe, G. E. and Moler, C.: *Computer solution of linear algebraic equations*, Prentice-Hall, New Jersey, 1967.
Fox, L.: *An introduction to numerical linear algebra*. Oxford University Press, N.Y., 1965.
Francis, J. G. F.: The Q R transformation, Parts I and II, *Comput. J.*, **4**, 1961–1962.
Franklin, J. N.: *Matrix theory*, Prentice-Hall, New Jersey, 1968.
Frazer, R. A., Duncan, W. J. and Collar, A. R.: *Elementary matrices and some applications to dynamics and differential equations*. Cambridge University Press, 1938.
Gantmacher, F.: *The theory of matrices*, Vol. 1 and 2. Chelsea, N.Y., 1960.
Goldstein, H.: *Classical mechanics*. Addison-Wesley, Reading, Mass., 1959.
Golub, G. H., Nash, S. and Van Loan, C.: A Hessenberg-Schur method for the problem AX + XB = C, *IEEE Trans. Automat. Contr.*, **AC-24**, No. 6, pp. 909–913, 1979.
Hanson, R. J. and Lawson, C. L.: Extensions and applications of the Householder algorithm for solving linear least-squares problems, *Math. Comput.*, **23**, pp. 787–812, 1969.
Hestnes, M. R. and Stiefel, E.: Methods of conjugate gradients for solving linear systems, *Nat. Bur. Standards J. Res.* **49**, pp. 409–436, 1952.
Hohn, F. E.: *Elementary matrix algebra*, Macmillan, N.Y., 1961.
Householder, A. S.: Lectures on numerical algebra. Notes on lectures given at the 1972 MAA summer seminar Williams College, Williamstown, Massachusetts, 1972.
Householder, A. S.: *The theory of matrices in numerical analysis*, Dover Publications, N.Y., 1975.
Kardestuncer, H.: *Elementary matrix analysis of structures*, McGraw-Hill, N.Y., 1974.
Kato, T.: *Perturbation theory for linear operators*. Springer-Verlag, N.Y., 1976.
Kreko, B.: *Linear programming*. Pitman, London, 1962.
Kreyszig, E.: *Advanced engineering mathematics*. John Wiley, N.Y., 1972.
Kublanovskaya, V. N.: On some algorithms for the solution of the complete eigenvalue problem, *Zh. Vych. Mat.* **1**, pp. 555–570, 1961.
Lancester, P.: *Lambda-matrices and vibrating systems*. Pergamon Press, London, 1966.
Lancester, P.: The theory of matrices. Lecture notes, the University of Calgary, Canada, 1963.
Landau, L. and Lifshitz, E.: *Mechanics*. Pergamon Press, London, 1969.
Laub., A. J.: A Schur method for solving Riccati equations, *IEEE Trans. Automat. Contr.*, **AC-24**, No. 6, pp. 913–921, 1979.
Levenberg, K.: A method for the solution of certain nonlinear problems in least-squares, *Quart. J. Appl. Math.*, **2**, pp. 164–168, 1944.

Macfarlane, A. G.: *Dynamical system models*. Harrap, London, 1970.
Marquardt, D.: An algorithm for least square estimation of nonlinear parameters, *J. Soc. Ind. Appl. Math.*, **11**, pp. 431–441, 1963.
Moler, C. B. and Stewart, G. W.: *An algorithm for the generalized eigenvalue problem* $Ax = \lambda Bx$. Tech. Rept CNA-32, Center for numerical analysis, Austin, Texas, 1971.
Myskis, A. D.: *Advanced mathematics for engineers*. Mir Publ., Moscow, 1975.
Nering, E. D.: *Linear algebra and matrix theory*, John Wiley, N.Y., 1963.
Ortega, J. and Rheinboldt, W.: *Iterative solution of nonlinear equations in several variables*. Academic Press, N.Y., 1970.
Papoulis, A.: Perturbations of the natural frequencies and eigenvectors of a network, *IEEE Trans. on Circuit Theory*, **CT-13**, No. 2, pp. 188–195, 1966.
Pease, M.: *Methods of matrix algebra*, Academic Press, N.Y., 1965.
Pestel, E. C. and Leckie, F. A.: *Matrix methods in elastomechanics*. McGraw-Hill, N.Y., 1963.
Peters, G. and Wilkinson, J. H.: The least-squares problem and pseudoinverses, *Comput. J.*, **13**, pp. 309–316, 1970.
Peters, G. and Wilkinson, J. H.: Ax = λBx and the generalized eigenvalue problem, *SIAM J. Numer. Anal.*, **7**, pp. 479–492, 1970.
Petrovsky, I. G.: *Lectures on the theory of integral equations*. Mir Publ., Moscow, 1971.
Pollard, J. H.: *Mathematical models for the growth of human populations*. Cambridge University Press, London, 1973.
Powell, M. J. D.: An efficient method of finding the minimum of a function of several variables without calculating derivatives, *Comput. J.* **7**, p. 155, 1964.
Rellich, F.: *Perturbation theory of eigenvalue problems*. Inst. of Math. Sciences, New York University, 1953.
Ribbans, J.: *Basic numerical analysis*, Book two. Intertext books, London, 1970.
Rohrer, R.: *Circuit theory, an introduction to the state variable approach*. McGraw-Hill, N.Y., 1970.
Rutishauser, H.: *Solution of eigenvalue problems with the L-R transformation*. Nat. Bur. Standards Appl. Math. Ser. **49**, pp. 47–81, 1958.
Salvadori, M. G. and Baron, M. L.: *Numerical methods in engineering*. Prentice-Hall of India, New Delhi, 1966.
Sobhy, M. I. and Deif., A. S.: The state space approach to first order perturbations in electrical networks, *IEEE Trans. on Circuits and Systems*, **CAS-22**, No. 10, 1975.
Sobhy, M. I. and Deif, A. S.: Application of the perturbation matrix in network design problems. *Electronics Letters*, **11**, No. 14, 1975.
Sobhy, M. I. and Deif, A. S.: The perturbation matrix and its applications in network design, *Int. Jour. Electronics*, **40**, No. 3, pp. 257–266, 1976.
Sokolnikoff, I.: *Advanced Calculus*. McGraw-Hill, N.Y., 1939.
Stoer, J. and Bulirsch, R.: *Introduction to numerical analysis*. Springer-Verlag, N.Y., 1980.
Temes, G. C. and Calahan, D. A.: Computer-aided network optimization, the state of the art, *Proc. IEEE*, **55**, No. 11, pp. 1832–1863, 1967.
Thomson, W. T.: *Theory of vibration with applications*. Prentice-Hall, New Jersey, 1972.
Thonstad, T.: *Education and manpower, theoretical models and empirical applications*. Oliver and Boyd, London, 1969.
Urquehart, N. S.: Computation of generalized inverse matrices which satisfy specified conditions, *SIAM Review*, **10**, pp. 216–218, 1968.
Van der Sluis, A.: Condition, equilibration and pivoting in linear algebraic systems, *Num. Math.*, **15**, pp. 74–86, 1970.
Wedderburn, J. H. M.: *Lectures on matrices*. American Mathematical Society Colloquium, **17**, 1934.
Wilkinson, J. H.: *The algebraic eigenvalue problem*, Clarendon Press, Oxford, 1965.

Wilkinson, J. H. and Reinsch, C.: *Handbook for automatic computation. Vol. 2, Linear algebra*. Springer-Verlag, Berlin, 1971.
Wolfe, P.: The secant method for simultaneous nonlinear equations, *Communs Ass. Comput. Mach.*, **2**, p. 12, 1959.
Young, D. M. and Gregory, R. T.: *A survey of numerical mathematics*, Vol. 2., Addison-Wesley, Reading, Mass., 1973.
Young, D. M.: *Iterative solution of large linear systems*. Academic Press, London, 1971.
Zadeh, L. and Desoer, C. A.: *Linear system theory*, McGraw-Hill, N.Y., 1963.
Zadeh, L. and Polak, E.: *System theory*. TATA McGraw-Hill, Bombay & New Delhi, 1969.

Index